New Surgery For Prostate Cancer

The da Vinci Robotic System

By

Sijo J. Parekattil, M.D.

Published by MWP Media
12231 Academy Rd. NE #301
Albuquerque, NM 87111
Cover designed by Rod Laverne

ISBN 0982878257
Printed in the United States of America

Table of Contents

Introduction

Hearing the word "cancer" when a patient is told that he has prostate cancer comes as a shock that often leaves him dazed. A man's response to the diagnosis that he has cancer is often shaped by awareness that his father, uncle or friend suffered from the illness and died prior to current treatments for prostate cancer.

There is good reason for concern when prostate cancer is detected. Keep in mind that it is the most common non-skin malignancy in men and except for lung cancer, often caused by smoking, it is responsible for more deaths than any other form of cancer.

Prostate Cancer Statistics

It is one of the most common cancers in men. About 140,000 new prostate cancers are diagnosed yearly in the U.S. and about 40,000 men die of prostate cancer each year. Estimates indicate that one in every ten men in the U.S. will develop prostate cancer in his lifetime.

It is interesting to note that in many older men, microscopic evidence of prostate cancer is discovered during autopsy. The American Cancer Society (ACS) estimates that nearly a quarter of a million men a year are diagnosed in the United States with prostate cancer, world-wide it is astounding the number. In the United

States, roughly 1 man in 6 will be diagnosed with prostate cancer during his lifetime, but only 1 man in 34 will die from it, though a little over 1.8 million men in the United States are survivors of prostate cancer due to improved treatment methods.

Today, there is much greater hope than ever before for achieving the goal of removing the cancer and retaining vital functions of the urinary system, as well as sexual function. The da Vinci robotic surgery system is the first to be approved by the FDA for general laparoscopic surgery using robotic methods.

Surgical robotics was little more than a medical curiosity until 1999, the year Intuitive Surgical introduced the *da Vinci*® Surgical System. Today, Intuitive Surgical is the global leader in the rapidly emerging field of robotic-assisted minimally invasive surgery.

The FDA has given its approval to the da Vinci robotic system for radical prostatectomy (prostate cancer surgery) and it is currently used in numerous hospitals across the country. Each da Vinci system is rigorously maintained, tested, and upgraded as necessary by *Intuitive Surgical,* the manufacturer of the system. It currently has the stamp of approval as an effective surgical treatment of prostate cancer.

Who may be eligible to undergo robotic prostate surgery?

- **Any patient with localized prostate cancer (has not spread outside the prostate)**
- **Any patient who can tolerate general anesthesia**

World Laparoscopy Hospital had this to say about the da Vinci robotic surgery:

"Robotic surgery is one of the leading medical advances in minimally access surgery. The da Vinci ® Surgical System is an enhanced computer which offers the surgeon high-definition 3D vision, greater precision and dexterity, and superior ergonomics, allowing the surgeon to perform beyond the limitations of human performance."

Chapter One
Prostate Cancer

As the male fetus develops in the womb, hormones cause the prostate gland to grow. It continues to enlarge in young men and progresses to full size in adult men and is roughly the size of a walnut.

This walnut size prostate gland is a vital part of the male reproductive system, as well as linked closely with the urinary system. The gland is located immediately below the bladder and in front of the rectum, surrounded by the urethra, the narrow tube that runs the length of the penis and carries both urine and semen out of the body. The small gland is flanked on either side by the seminal vesicles, a pair of pouch-like glands that contribute secretions to the semen. Next to the seminal vesicles run the two vas deferens. These are tubes that carry sperm from the testicles, where sperm is manufactured, and where testosterone, a male sex hormone that controls the prostate's growth, is made. Its primary function is to produce a substance that makes up part of the milky fluid called semen.

The entire process delivers sperm through the penis during ejaculation for reproductive purposes with the sides of the gland engulfed with essential nerves, which control erectile function.

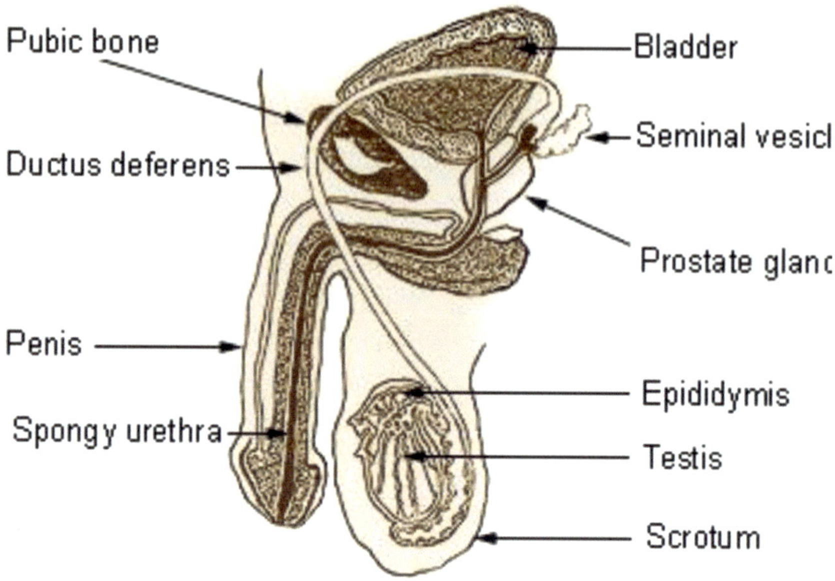

Male Reproductive System

If the urologist informs a patient that his biopsy test clearly indicates that he has prostate cancer, and that it is localized for the time being in the prostate gland, that patient must act in concert with his doctor on a sound treatment course. Since there are no noticeable symptoms of prostate cancer while in its most treatable form, the option of surgery to remove the gland is often considered. The malignancy must be caught before a series of tests indicate a more advanced stage and the spread to other areas of the body. Early symptoms may include difficult or frequent urination or blood in the urine. At this stage, when the patient learns of his condition that cancer has been detected, then most likely the treatment entails some medical treatment to halt the spread.

When a patient learns that he has prostate cancer, most surgeons will explain the options available, other than surgical removal of the gland. Yet, knowledgeable surgeons know that one very effective, proven method to be free of the possibility of the cancer spreading is surgery, provided the cancer, in fact, has not metastasized, and is confined to the prostate gland.

Types of Surgery

Often, patients are advised to discuss treatment options with two cancer specialists – a urologist and a radiation oncologist. These specialists usually devote part of their practice to working with men suffering from prostate cancer. It is wise for any patient to get the best advice possible of options other than surgery for prostate cancer. Among the options for treatment are radiation therapy, including brachytherapy and intensity-modulated radiation therapy, new trials of chemotherapy and vaccine treatments.

If the patient is a good candidate for prostate cancer surgery: in good health, tumor confined to the prostate gland, under 70 years of age (approximate age range), and expected to live another 10 years, the surgical treatment of prostate cancer is acceptable. If tests indicate that the prostate cancer is confined to the prostate, then surgery to remove the gland is often a viable course of action.

Confronted with the possibility of surgery, there can be a good deal of patient anxiety. Most concerns are related to surgical pain, post surgical recovery and side effects. The patient wants to know if his body will be impaired or compromised in any way. How much pain will there be? What kind of side effects would there be? Would there be impotence or incontinence?

Fortunately, less invasive surgical options are available to many patients facing urologic surgery. A new category of surgery, introduced with the development of the *da Vinci*® Surgical System, is performed by a growing number of surgeons throughout the country. These new approaches utilize the latest in surgical and robotic technology. Robotic assisted laparoscopic prostatectomy is an effective method of prostate gland removal. The goal is to try to optimize the removal of the prostate without leaving any cancer behind, minimize any injury to the surrounding nerves affecting sexual function, and minimize any damage to the urinary sphincter to prevent any problems with urinary control.

The standard option for the surgical removal of the prostate gland in the past has been open surgery (referred to as the Gold Standard). This is called open radical prostatectomy. There is little question that surgery to remove the cancerous prostate gland is an effective method of treatment. How that surgical removal is performed becomes the issue of concern.

One of the greatest drawbacks to open prostate surgery is the requirement for a large incision in the abdomen. The *open* surgery approach has operative and post operative side effects that have to be considered. It leaves a large scar, has a risk of excessive blood loss during surgery, may require an extended hospital stay, and may cause considerable pain during the first phases of recovery.

A more recent advanced method of surgical removal of the prostate gland is the less invasive laparoscopic approach. This procedure is performed with several small incisions or ports and laparoscopic instruments. These instruments are inserted through the ports while the abdomen is inflated with gas (carbon dioxide). A tiny camera is inserted through one of the incisions/ports for the surgeon to view and manipulate the instruments. This method is quite tedious and requires extensive training for the surgeon. However, the minimally invasive approach offers significant advantages over open surgery.

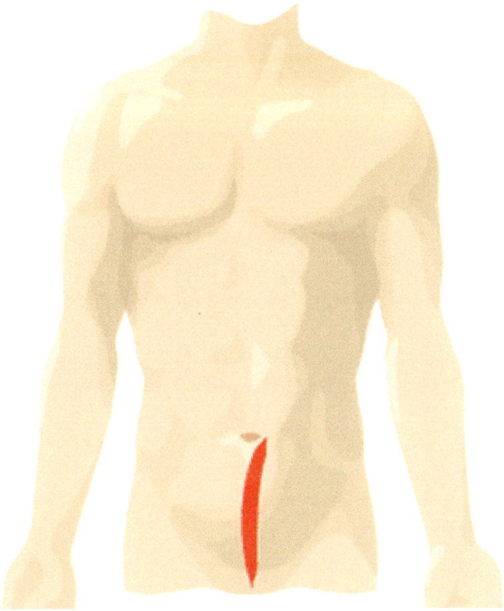

Open Prostate Surgery with a large incision in the abdomen

The word laparoscopy means to look inside the abdomen with a special camera or scope. This surgery with the aid of cameras is also called keyhole surgery. As mentioned, traditional open surgery requires a long incision (cut) down the center of the abdomen and a lengthy recovery period. Laparoscopic surgery eliminates the need for a single, large incision. The result of laparoscopic surgery is less pain, small scars and faster recovery.

Laparoscopic prostate surgery involves five small (5 to 10 millimeters) incisions with one placed near the navel and two each on both sides of the lower abdomen. Carbon dioxide is passed into the abdominal cavity through the incision near the navel with a small tube. The inflated abdominal wall offers the surgeon an excellent view of the abdominal cavity once the laparoscope is in place. The laparoscopic camera provides a digital, high definition view of the prostate and the delicate surrounding structures that enables the surgeon to perform the procedure carefully.

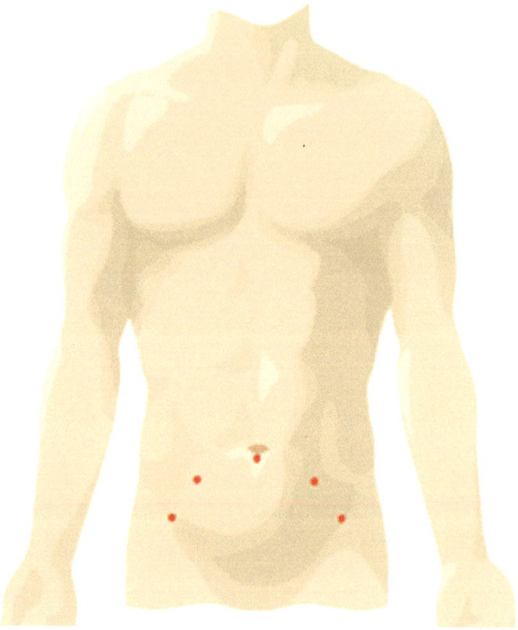

Laparoscopic Prostate Surgery Small Incisions

Men who undergo the laparoscopic technique have less blood loss, less need for pain medication, a shorter hospital stay, quicker return to regular meals and activities, early removal of urethral catheters (which are tubes inserted through the penis to drain urine from the bladder), and a quicker recovery. In short, laparoscopy treats the prostate cancer as effectively as surgeries performing the open procedure, but with advantages as listed above.

However, the laparoscopic surgical approach to prostate removal offers only a two dimensional view of the surgical site. A more ergonomic and somewhat futurist surgical procedure is the robotic assisted prostatectomy using the *da Vinci Robotic System.* This has taken laparoscopy to a new level of treatment for prostate surgery. It offers the advantages of traditional laparoscopic surgery, but also provides an immersive 3 dimensional view of the surgery with an ergonomic, stable, easy to use surgical platform.

The da Vinci system not only reduces operative duration for complex surgical procedures, but also makes it easier for surgeons to perform complex laparoscopic procedures. The hope is that these advantages will translate to more effective methods to have the erectile nerve sparing and urinary muscle preservation during prostatectomy.

Who qualifies as a candidate?

The following are requirements for the robotic surgery:

- **Prostate cancer must be clinically confined to the gland**
- **Prostate cancer in the early stages. This means the tests indicate that the cancer does not extend beyond the capsule**
- **The candidate must be sufficiently healthy to undergo surgery and expected to live for 10 years**

More than 2,500 surgeons from every corner of the globe have been trained in da Vinci Robotic Surgery and the number of procedures and cases being performed with this system are rising rapidly. Last year (in 2009), almost 70-80% of all prostatectomies were done using the robotic system.

Chapter Two
The da Vinci Surgical System for Prostate Cancer

Surgical management of prostate cancer is at a new level of efficiency and precision. The latest use of robotic-assisted surgery allows the surgeon to perform complex surgery using small incisions and have most patients experience a recovery period of less than 48 hours.

Ryan Rhodes, Senior Director of Marketing for *Intuitive Surgical*, claims that there have been more than 800 previous studies on robot-assisted prostate surgery undertaken. "The overwhelming majority of these show good results, for cancer treatment, urinary continence and sexual function."

The da Vinci Robotic Surgery System offers the patient tiny incisions, minimum blood loss and a potential speedy recovery. It also employs robotic laparoscopic instrumentation that provides the skilled surgeon with unparalleled precision and stability during the surgical procedure. The new high definition stereoscopic cameras offer the skilled surgeon a magnified immersive 3-dimensional experience while seated at a console.

The da Vinci system comprises of three components: a master surgeon console; a bedside robotic surgical manipulator; and a control tower. During surgery, the surgeon sits at the master surgeon console and controls the robotic surgical manipulator. With the new da Vinci system, up to two surgeon consoles may be used so that dual surgeons may operate simultaneously if necessary. However, usually there is only a single master surgeon console and one surgeon.

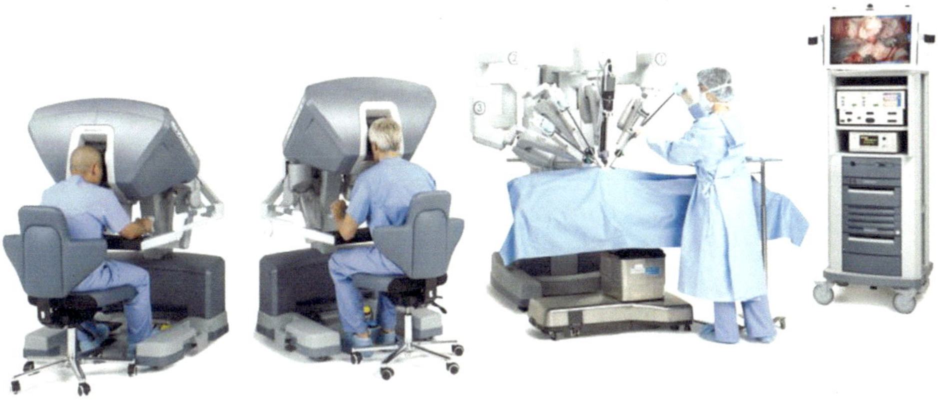

Operating theater during the da Vinci robotic surgical procedure

©2010 Intuitive Surgical, Inc.

The surgeon console consists of a stereoscopic eyepiece that allows the surgeon to view an immersive three-dimensional image of the surgical field, two hand controls that allow control of the robot's arms, and a number of foot pedals that allow the surgeon to select between control of the camera and instruments. The foot pedals also allow the surgeon to activate and deactivate certain tools during the procedure. The control tower unit contains the computer that integrates the images from the cameras, equipment for the carbon dioxide insufflation of the abdomen, the light source for the camera, and a monitor that allows the other operating room staff to observe the surgeon's view.

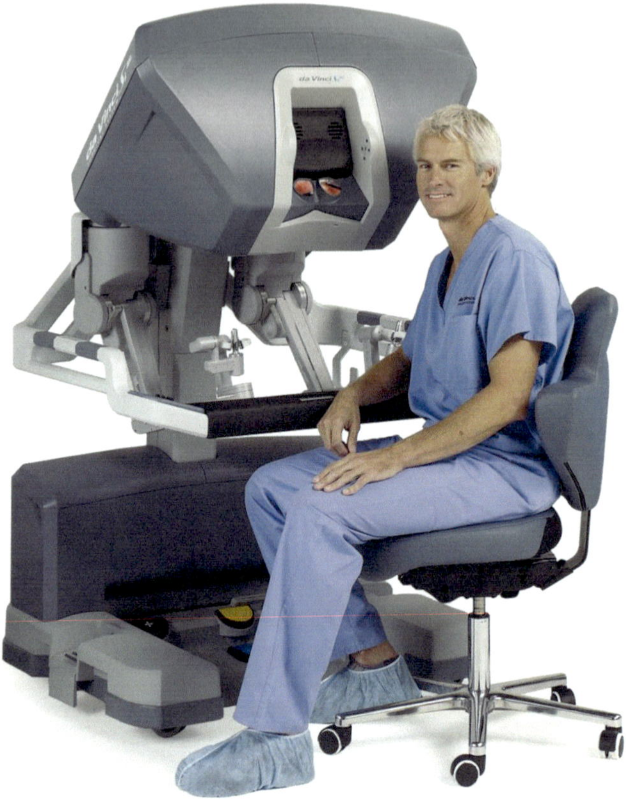

Surgeon during surgery, at the da Vinci master surgeon console

©2010 Intuitive Surgical, Inc.

The da Vinci robotic system is a very futuristic system that allows the surgeon to also incorporate views from other instruments, such as ultrasound, high magnification accessory cameras with additional views, or MRI fusion imaging with the images from the operative field. When looking inside the stereoscopic eyepiece – the surgeon can either toggle these views or watch them simultaneously next to each other. This almost mimics the cockpit of an advanced fighter jet. During surgery, the surgeon is seated at the console where he/she can perform the surgery with all the imaging and instrumentation at the surgeon's fingertips in an ergonomic platform. The console actually conforms to the surgeon's posture and can be adjusted with six degrees of freedom to accommodate every surgeon's individual needs.

The console offers a platform where the surgeon can control multiple instruments. By moving the hand controls together like a steering wheel and simultaneously pressing the camera foot pedal, the surgeon can move the camera with seven degrees of freedom (and two axial degrees of rotation). Once the camera foot pedal is released, the hand controls control the left and right robotic arms respectively. Any instrument held by the robotic arms can be manipulated with seven degrees of freedom as well. Either hand control can also be toggled to manipulate an additional 3rd robotic instrument arm during the surgery. The robotic system, in the hands of an experienced robotic surgeon, can be an incredible tool in performing radical prostatectomy.

The da Vinci 3D High Definition Optics

The da Vinci robotic camera consists of two high definition cameras. This is much like the human eye that offers a true 3D color image. The images are visible to the surgeon seated at the surgeon console by viewing through the twin high resolution, high framerate eyepieces.

The surgeon's view inside the abdomen is actually an endoscope picture with magnification of up to 12 to 15x. By viewing through the stereoscopic eyepieces, the two separate high definition images are merged to create true 3-dimensional immersive vision for the surgeon. This experience is similar to that when one watches a 3D IMAX theatre movie. It provides true depth perception to the surgeon that is usually not available with standard laparoscopy. This bigger than life view that the camera captures of the ongoing procedure is one of the most advanced vision systems available today.

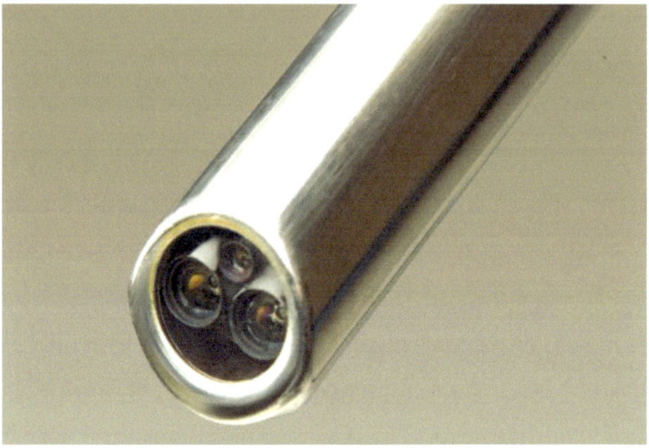

This illustration shows the tiny cameras used in robotic surgery.

©2010 Intuitive Surgical, Inc.

The camera lens tip can either be straight or angled. The angled lens allows the surgeon to view around objects and is sometimes used in certain procedures to enhance the viewing ability of the surgeon.

The central robotic arm on the bedside robotic instrument manipulator positions the camera and lighting exactly where the surgeon directs it. It offers a stable visual view while the surgeon is operating with the other robotic instrument arms. The camera can be positioned within inches from the prostate and can expose the area of surgery with simultaneous clarity and stability that is hard to achieve with standard laparoscopy. This enhanced vision with ergonomic instrument control allows the surgeon to perform complex surgical laparoscopic maneuvers with greater ease than if the surgeon were using straight handheld laparoscopic instruments (as in standard laparoscopy).

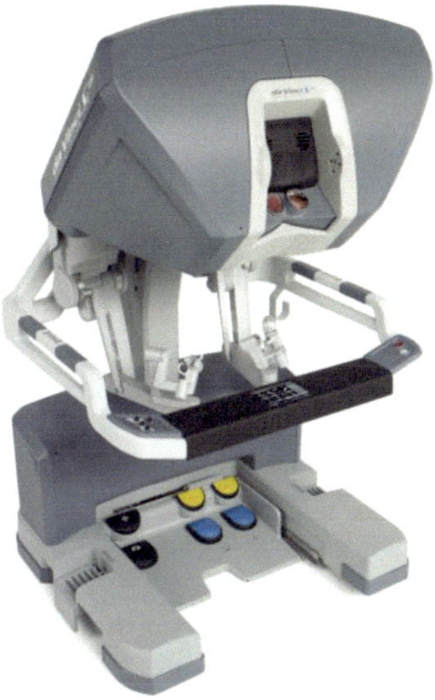

The high resolution, high frame-rate eyepieces for viewing 3D images with 12 to 15x magnification.

©2010 Intuitive Surgical, Inc.

As with standard laparoscopic surgery, the robotic approach requires insufflation of the abdomen with carbon dioxide gas to elevate the abdominal wall and allow an unobstructed view of the internal organs. At the conclusion of surgery, the gas is released through the ports and the body absorbs any residual CO2. The patient may feel a little bloated and may sometimes have some shoulder pain (due to irritation of the diaphragm from the gas). The gas is non-flammable, which is essential to guard against burns from the use of electrosurgical instruments.

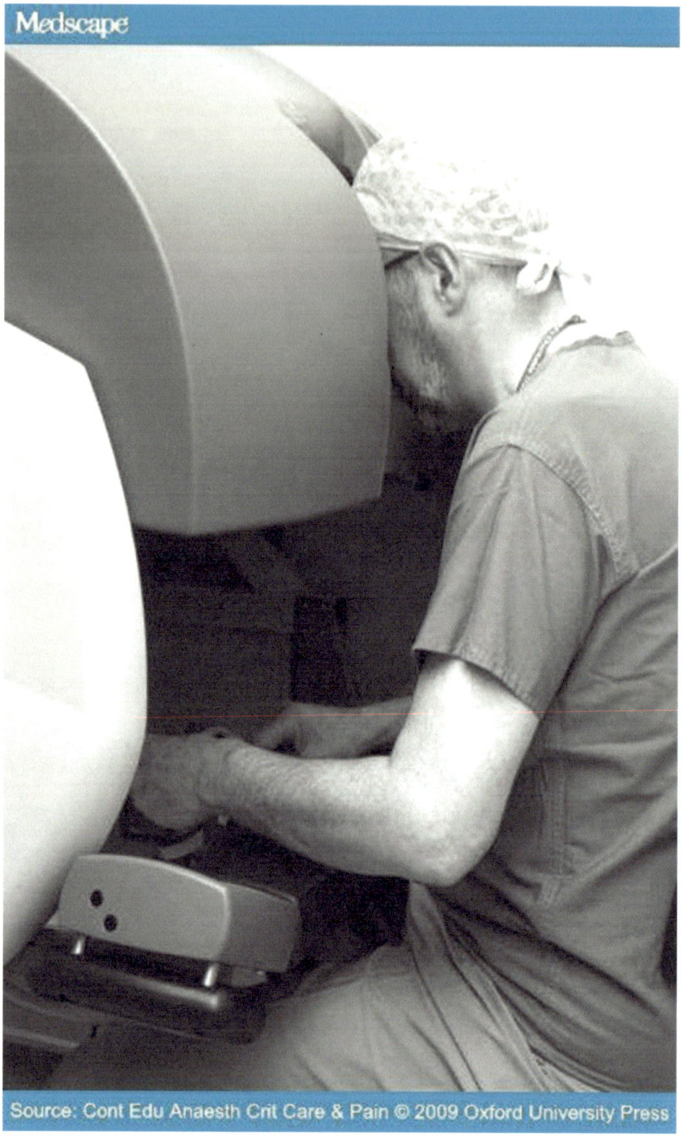

Bedside Robotic Surgical Manipulator

The part of the robotic system that is actually holding the instruments next to the patient is called the robotic surgical manipulator. It has four arms; one holds a camera and the other three hold interchangeable surgical instruments that are inserted into the patient through small incisions called ports.

The arms of the robotic surgical manipulator are capable of accurately applying force and have a wide range of motion with seven degrees of freedom. This motion can range from small movements, allowing for precision work, to large movements, such as moving bulky tissues or organs that may be in the path.

The surgeon remains at the console while a surgical assistant is at bedside. The assistant helps in the insertion of the tiny instruments through the ports (incisions) and in changing the various instruments.

The arms of the robot can hold and manipulate a host of instruments such as the camera, as well as tools to accomplish the operation. One surgeon stated, "This system gives the surgeon another pair of arms to work with; it's like having four arms working simultaneously."

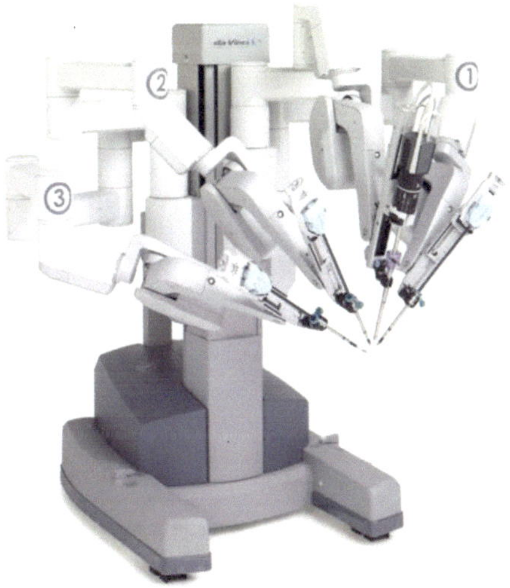

The robotic surgical manipulator has four arms and holds three surgical instruments and a telescopic camera. The hand-like surgical instruments move with seven degrees of freedom and two degrees of axial rotation.

Manipulation of da Vinci Surgical Instruments

Standard laparoscopic instruments are manipulated counter-intuitively or 'backwards.' During a traditional laparoscopic surgical procedure, while the surgeon is pushing down on one end of the instrument, the other end goes up, and when the surgeon pushes right, it causes a left movement as the other end goes up.

The da Vinci robotic system corrects for this reversal. The surgeon's hand movements are identically mirrored in the robotic instruments. If the surgeon's hand turns left, the robotic hand turns left. With the aid of computer correction, the robotic instrument motion mimics the surgeon hand motion precisely. This makes performing complex laparoscopic maneuvers much easier.

The surgeon at the console can perform micro precise dissection with these mechanical advantages. Studies have shown that the robotic system reduces surgeon fatigue due to the fact that the surgeon is manipulating the instruments seated in an ergonomic position with little force on the hands.

Full Range of Hand Motion

The robotic approach has an additional advantage over standard laparoscopy in providing articulation of the instrument tips. This allows the instruments to not only merely open and close, but also to fully turn and twist. This gives the surgeon at the console a more natural mimicry of the movements of the human hand and wrist. The instrument tips are also miniaturized. Most of the jaws of these tiny tools are similar or shorter in length than fingernails. This allows very small and precise movements to carefully dissect out the prostate.

The robotic system translates a surgeon's hand movements exactly during surgery. If the surgeon turns his/her wrist right, the articulated robotic wrist turns right. If the wrist goes up, the robotic instrument shifts up. The system also allows the surgeon to scale the motion. For example, the system can be set to 4:1 scaling – so for every 4 inches moved by the surgeon's hand, the robotic instrument will only move 1 inch. This allows the surgeon to perform very delicate micro surgery rather easily with minimal fatigue.

All surgeons have some degree of hand tremor. The robotic system also has the ability to filter out tremors. This offers the surgeon more precision than possible with standard laparoscopy.

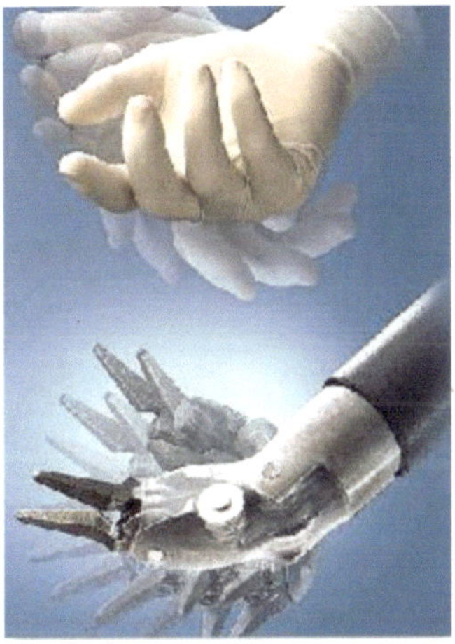

The proprietary *EndoWrist*® Instruments powered by state-of-the-art robotic technology offer hand movements that are scaled, filtered and seamlessly translated into precise movements.

During surgery, the surgeon is seated at the console. There are two hand controls within the console. The surgeon/operator places his/her index finger and thumb in the ring loops on each hand control. Opening and closing instrument tips is performed by opening or closing the ring loops on the hand control. The amount of pressure applied when closing these loops corresponds to the amount of pressure applied by the jaws of the robotic instrument when grasping tissues or suture.

Specialized technology simultaneously transfers the surgeon's exact hand movements at the console to the extremely precise micro-surgical movements of the robotic pincers ("hands") inside the patient's abdomen. Through the incisions, the surgeon can rotate the instruments with 7 degrees of freedom.

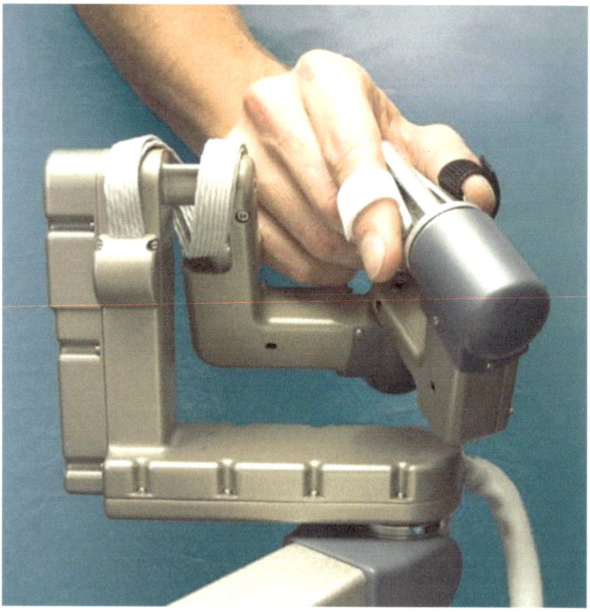

Robotic hand control in the surgeon console

©2010 Intuitive Surgical, Inc.

There are those who insist that a surgeon needs to use fingers to gain a sense of touch and feel in order to effectively remove cancer. However, the enhanced visual experience and precision fine movements can compensate for some of this lack of tactile feedback. Generally speaking, robotic prostatectomy patients are patients that have organ confined disease, and the tactile sensation of cancer in the prostate may not be as critical since the disease is primarily within the prostate. There are cases of high grade prostate cancer, where open approaches may offer this advantage. However, there is more evidence to support alternative approaches, such as radiation treatment, for patients with advanced disease and the role of surgery in general in these high risk patients may be diminished.

During open surgery, the prostate gland is in a position that makes it slightly difficult to visualize some of the critical structures next to the prostate. This is what led to the use of surgical magnification loupes (special magnifying glasses worn by the surgeon) during the surgery. The robotic approach could be merely thought of as a further extension of this concept – enhancing the visual experience for the surgeon.

Because the prostate is located underneath and deeply embedded behind the pubic bone area - it tends to be in an area that is deep and slightly difficult to visualize easily. Therefore, surgeons rely on the use of surgical magnification loupes and the tactile feeling around the prostate to achieve successful surgery. Highly experienced open surgeons have learned how to augment all their senses to achieve good results with the open technique.

Robotic surgery offers advantages in terms of offering a highly magnified 3-D view underneath the pubic bone or any other difficult to view location, deep in the pelvis and near the vital structures. The surgeon at the console can clearly see what must be removed. As one surgeon put it, “the surgeon is virtually placed inside the patient's body“.

The da Vinci surgical instruments, though at first glance similar to standard laparoscopic instruments, have an advantage: they can be articulated. This gives the instruments not only the ability to open and close, but to fully turn and twist. These surgical instruments are designed with jointed wrists, which provide freedom of movement in multiple directions (7 degrees of freedom).

It is as if a human hand were twisting and turning in motion. Yet, unlike the surgeon's hand, these instruments are considerably smaller. Some of the "jaws" of the tools are similar, and with some, shorter in length than a fingernail. These amazingly accurate and tiny instruments make small and precise movements to carefully dissect out the prostate.

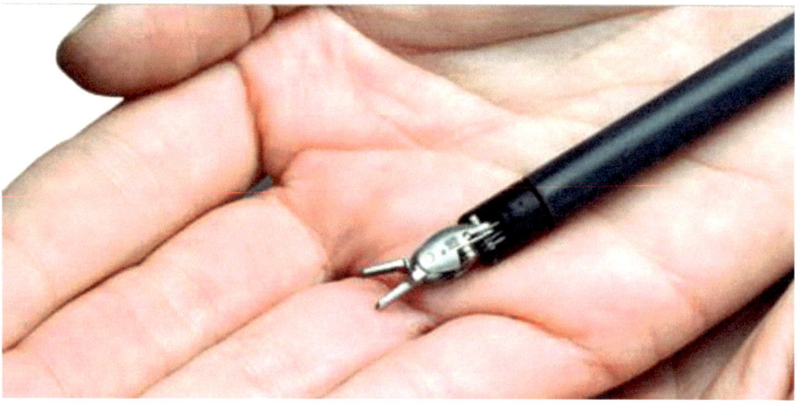

The da Vinci System's Small Hand-Movement Tools

©2010 Intuitive Surgical, Inc.

How Robot-Assisted Surgery Works

As with minimally invasive surgery (MIS), the surgeon inserts miniature surgical instruments and a camera into small incisions in the patient. Unlike MIS, the surgeon controls the movement of the instruments and camera at a robotic console in the operating room. The console is equipped with a special three-dimensional monitor and intuitive hand controls. The surgeon console contains the master controls that the surgeon uses to manipulate the EndoWrist Instruments.

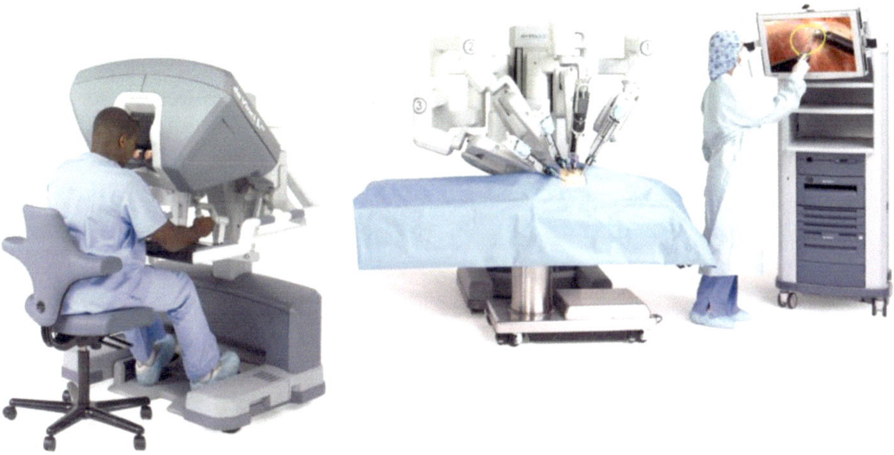

Typical Setup Preparing for Robotic Surgery

©2010 Intuitive Surgical, Inc.

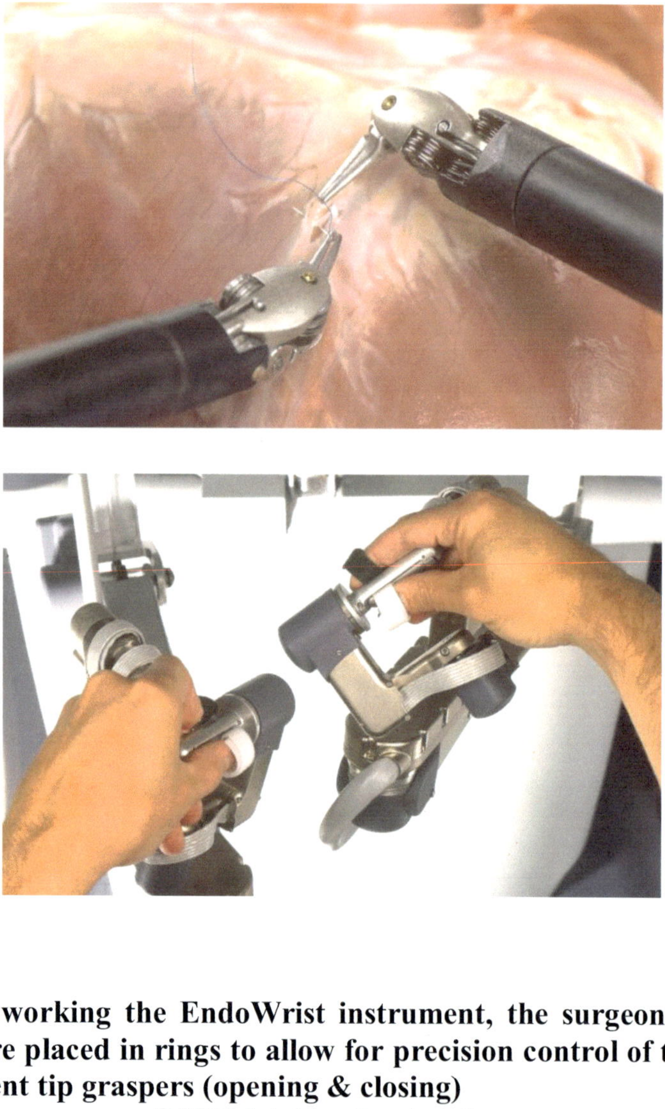

While working the EndoWrist instrument, the surgeon's fingers are placed in rings to allow for precision control of the instrument tip graspers (opening & closing)

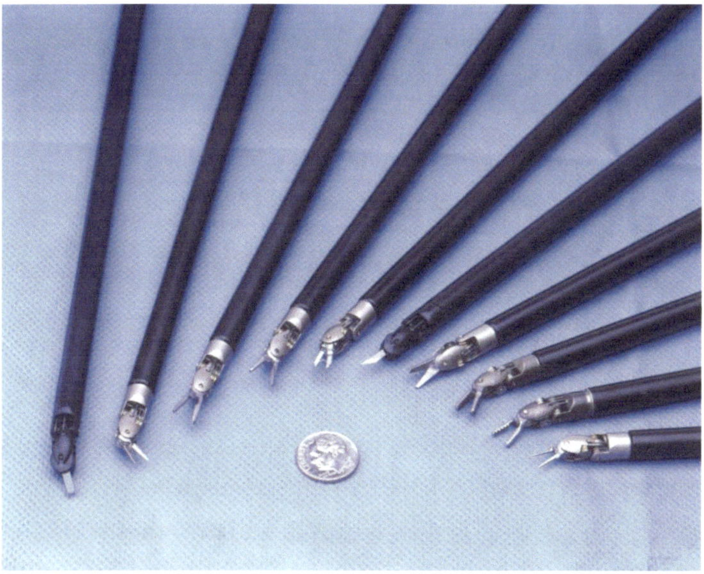

The specialized technology of the da Vinci System transfers the surgeon's exact hand movements at the console to the micro-surgical movements of the three robotic instruments ("hands") inside the patient's body. The surgeon can manipulate the instrument tips with 7 degrees of freedom.

Minimizing Blood Loss with the da Vinci System

Robotic surgery with an experienced robotic surgeon offers small incisions and fine dissection that allows for reduced blood loss for the patient. The surgeon, through the magnified images from the camera, has a much greater detail of vision and can work around nerves and blood vessels, which helps minimize bleeding.

Blood Loss during Surgery

Open prostate surgery on average has greater blood loss than robotic laparoscopic prostatectomy. The average blood loss in open surgery is around 300 to 1000 cc's (16 to 32 ounces). The average blood loss in robotic laparoscopic surgery is 150 to 250 cc's (3 to 8 ounces). Possible blood transfusion with robotic prostatectomy is less than 1%, while it can be as high as 10%-15% in open surgery.

Training and Surgical Skill

Robotic laparoscopic prostatectomy requires the back-up skills by the surgeon to convert to laparoscopic or open surgery should the need arise. There is a learning curve with robotic prostatectomy and experience helps. Surgeons who have performed over 250 cases are considered to have sufficient experience to be considered at a good level of performance. Whether the surgeon has had fellowship training in robotic surgery is also an added advantage. This is specialized training beyond the general urology residency training where the surgeon has received further specialized training in robotic surgery. Experience varies with each surgeon, but it tends to be a measurement of proficiency that offers the best outcomes with use of the da Vinci robotic system.

Unlikely System Failure

If, for some reason, the surgeon senses that there is a malfunction, or detects that it is no longer safe to continue with the robot system during surgery, the da Vinci system will be withdrawn and the surgery will proceed either with the laparoscopic method or traditional open radical prostatectomy. Thankfully, system failures that require conversion to other methods are exceedingly rare.

Qualifications in a Robotic Surgeon

The best and most qualified surgeon includes the following:

- **Experience: must have performed 250 robotic prostatectomies with good results**
- **Willing to share or publish personal results of the robotic prostatectomies performed**
- **Compassion in the patient relationship and personal attention to the patient**
- **Fellowship trained in minimally invasive surgery and/or oncologic (cancer) surgery**
- **Acceptable complication rates**
- **Must have a skilled assistant surgeon and skilled operating room team**

The Safety of the da Vinci Robotic Surgery for Radical Prostatectomy

The system is FDA approved for robotic radical prostatectomy with a qualified robotic surgeon. It is routinely employed in many medical facilities around the world. Each da Vinci system is carefully tested, and upgraded as necessary by Intuitive Surgical.

Chapter Three
Urinary and Sexual Side Effects: Nerve Sparing

The three key aspects of great concern to most patients are: cancer cure, preservation of urinary control, and sexual survival. This is known as the golden "Trifecta". Patients anticipating prostate surgery want to know if the nerves near the prostate can be spared to offer them the same quality of life as before surgery. There is great hope with today's new and very precise manner of removing the prostate gland that better methods of preserving the nerves may be achieved.

Loss of Urinary Control (Incontinence) after Surgery

Urinary problems are common both before and after surgery for prostate cancer. Prior to treatment, a patient may experience trouble urinating due to an enlarged prostate. There are patients who prior to surgery have difficulty with the flow of urine or may have dribbling. In these cases, removal of the prostate may also benefit the patient in relieving some of these symptoms.

After surgery, many patients find that they have a very real problem of urinary incontinence. This is the medical term for no control of urine. Patients may leak urine when they are not aware, when performing activities, or continuously. Usually, this incontinence is temporary, meaning it may last a few weeks to months. However, some patients will always have some mild degree of incontinence (may leak a few drops in their underwear with straining).

Recovery of urinary control is dependent on many factors. Some of these factors are out of the surgeon's control, such as a patient's other medical conditions and the ability of the patient's urinary sphincter to recover after surgery.

Technique of Preserving Urinary Control

Daily, the kidneys empty roughly 3 pints of urine into the bladder. The wall of the bladder is made of stretchy muscle which allows it to stretch like a balloon as it fills with urine. This is normal. Most people have no urge to pass urine when there is about a quarter of a pint of urine in the bladder. The normal adult bladder can hold about 1 pint of urine.

There is a ring of muscle at the 'neck' of the bladder that extends through the prostate all the way to the urinary sphincter (the muscle that contracts and relaxes to allow a person to urinate). This entire muscle complex acts much like a tap, keeping the urine in. This muscle complex is strengthened greatly by the muscles and ligaments of the pelvic floor (that stretch backwards from the pubic bone to the backbone). The pelvic floor muscles are like a trampoline. The pelvic floor separates the pelvic cavity above from the perineal region below.

When the bladder contains about ½ pint of urine, urges in the form of signals to the brain indicate a need to urinate. When it is appropriate to urinate and the brain gives the signal, the bladder muscle stops stretching and begins to contract, squeezing the urine out. At the same time, the bladder neck, urinary sphincter and pelvic floor muscles relax, allowing the urine to pass down a tube, called the urethra, to the outside. In men, the tube is long and passes through the prostate gland before it reaches the penis to release the urine through the hole in the head of the penis.

Since a portion of the urethra passes through the prostate and is removed during surgery with the prostate, attention to detail is important. During robotic surgery, the amount of bladder neck musculature that is spared, the length of urethra that is able to be preserved at the tip (apex) of the prostate, and avoidance of injury to the pelvic floor muscle fibers all come into play.

The surgeon may study the prostate ultrasound pictures or cystoscopy images (taking a look inside the bladder with a scope) to create a pre-operative blueprint of how he/she is going to approach the bladder neck and urethra. Having this blueprint lets the surgeon plan out and contour the surgical movements so as to preserve the most bladder neck muscle possible. Some patients may have a protrusion of the prostate into the bladder at the bladder neck – called a large median lobe. This can make the surgery more tedious technically.

During real time surgery, the doctor will be cognizant of the pre-surgical blue print as he/she is seated at the console performing the very delicate cutting of tissue surrounding the prostate gland. It is during this crucial period that experience and skill count greatly.

Urinary Control Rehabilitation after Surgery

Many centers now utilize well established physical therapy and behavioral therapy protocols to help in the recovery of urinary control after surgery.

Physical therapy incorporates modalities to strengthen the pelvic floor and urinary sphincter muscles after surgery. Techniques such as biofeedback may be used: a technique where small electrodes are placed in the pelvic floor muscles to help teach patients how to correctly squeeze and strengthen certain muscles.

Behavioral therapy incorporates changes in drinking habits, eating habits and urinary habits to maximize urinary control.

All these techniques take time and patience, and the patient has to be prepared to work with the physical therapy team over weeks to months to gain the most benefit. Fortunately, many patients currently achieve a greater than 90% rate of complete urinary control within months after surgery and many do not require much rehabilitation.

Sexual Dysfunction after Surgery

The ability to have an erection is especially important for most men considering surgery. Facing erectile dysfunction (ED) as a result of prostate cancer treatment can be devastating, but there are a number of new intra-operative techniques during surgery and a number of successful rehabilitation techniques after surgery to minimize ED. Despite these advances, it is important to note that some men will not experience a *spontaneous* erection after surgery. However, there are methods to compensate for lack of spontaneous erection and this is not something to fear – it just requires some patience and effort on the part of the patient.

All men who have surgery will also cease to have an ejaculate. These men may have erections and orgasms, but since the prostate and seminal vesicles (responsible for producing the ejaculate) are removed, there will no longer be any ejaculate fluid that comes out at orgasm. Some men do get a small amount of fluid that comes out with orgasm after surgery – this fluid is just a small collection of some of the excretions from the urethral glands and a little bit of urine.

Nerve Preservation during Surgery to Minimize ED

Dr. Patrick Walsh, a pioneer in open radical prostate surgery from Johns Hopkins, first described the location of the nerves adjacent to the bottom portion of the prostate that are critical to men's erectile functions.

Since the publication of his landmark study, other data has emerged detailing the location of additional nerve fibers present in the tissue extending up the sides of the prostate gland. New techniques of athermally (without using any cautery) dissecting these nerves, using cooling methods and hydrodissection (using water to gently separate planes) techniques have been developed to try to further optimize outcomes.

There are bundles of nerves on either side of the prostate gland and these nerves for the most part are responsible for the ability to have an erection. During da Vinci prostate surgery, if these nerves appear to be cancer-free, the surgeon will not remove them when removing the prostate gland. Preservation of these nerve fibers has produced improved sexual function outcomes in the hands of skilled surgeons. With these techniques, surgeons have seen significant improvements in erectile function recovery.

Keep in mind that men can still have erections along with a normal sex drive, sensation and orgasm even if only a portion of the nerves are saved and thus it is critical to preserve as much of these nerves as safely possible without leaving any cancer behind. The prostate can be thought of as an onion, with the outer layers carrying the nerves – the goal is to carefully peel the outer layer off the prostate without cutting into the prostate and without damaging the delicate outer layer.

Experience provides an advantage for surgeons when performing this delicate nerve sparing. Outcomes tend to be better at high volume centers that perform large numbers of these procedures compared to small volume centers. One study of medical centers that perform high volume robotic surgery indicated an over 80 percent potency rate at 18 months post surgery.

Nerve Sparing Surgery

With the da Vinci robotic surgery system, nerve sparing surgery can be done with a high degree of accuracy:

- **The field of surgery is magnified 12-15 times, making it easier to recognize the small, delicate neurovascular bundles.**
- **The tips of the microscopic instrument can rotate, bend, turn, swivel, like the human wrist. This is called the *Endowrist Technology*.**
- **The wrist like movement around the prostate gland gives the surgeon dissecting the neurovascular bundles greater freedom.**

Erectile/Sexual Function Rehabilitation after Surgery

Many surgeons today have created detailed programs of pre-surgical preparation and post-operative rehabilitation to maximize their patient's ability to return to their pre-surgery level of sexual function and to be able to achieve this as rapidly as possible.

A formalized rehabilitation program works to maximize tissue oxygenation, while reducing scarring that can slow or prevent full recovery. It needs to be noted that not all patients return to their pre-surgical level of sexual function, but rehabilitation is an essential assist to most patients and increases the chances of reaching an acceptable level of sexual function.

Factors for Sexual Recovery after Surgery

- **Patients are more likely to regain potency, if potent before surgery.**
- **The skill and experience of the surgeon in sparing the nerves during surgery is a major factor in potency.**
- **The patient's unique physiological and psychological make-up makes a difference.**
- **The patient's psychological make-up can make it more difficult to regain potency after surgery.**

A Patient's Comments on Sexual Intercourse after Surgery

"It is now 9 months since my surgery; I have had a few nocturnal erections. Whenever I have unassisted erections, they are not 'stuffable' (yet). My best erections occur when I use Viagra along with the pump. I have been taking 25 mg of Viagra every other night or so, to encourage nocturnal erections – with limited success. I have not yet tried Muse or injections."

Based on my experience, there are several things I want to say about intimacy after RP.

First and foremost, as I stated earlier, sex will never be the same. You have to accept that and move on.

"I miss the ejaculate and the feeling that went along with it. But I realize that with my prostate gone, the ejaculate will never come back. Those days are over."

"I miss natural (i.e., unassisted) erections – i.e., erections that occur not because of Viagra or the pump but because my partner is beside me and I am "turned on." I hope those natural erections return -- and soon! I don't like the idea of having to 'plan' and 'prepare for' sex. That certainly spoils the spontaneity of sex."

"My orgasms are more intense ('sharper,' if you will), but I would rather have my prostate back (as long as it would be non-cancerous). Other men also report their experience of orgasm changed as a result of surgery. Some report their orgasms are BETTER; others report they are not as good as before."

"I also have learned that intimacy and pleasure do not require penetration. Kissing, caressing, and holding can be sensuous and enjoyable."

"Within three or four weeks of surgery, my libido — 'the urge to merge' — was back to normal. I have had no decrease whatsoever in my desire to have sex. After RP, I have undergone no further treatment — no radiation, no hormonal treatment, no chemotherapy."

"Don't be afraid to try the pump, Viagra, Muse, and injections. See what works best for you. I have yet to try the latter two, but I haven't ruled them out. I want to remind you that each person's experience is different. I have shared my experience, but I don't know what yours will be. With time, patience, and understanding, sex after prostate surgery can be very satisfying."

Chapter Four
Pre-Operative Preparation

Prior to surgery, a patient receives an instructional sheet from the surgeon's office that will detail the steps to be taken in preparation for da Vinci Robotic Surgery.

Medications

Medications You MAY Take Day of Surgery	Medications You May NOT Take Day of Surgery
Take half of your insulin dose	Water Pills
Heart Pills	Diabetic Pills
Blood Pressure Pills	Aspirin (stop minimum of 10 days prior to surgery)
Seizure Pills	Coumadin (generally stop 7 days prior to surgery-this must be cleared by the prescribing physician)

- Bring all medications in their labeled containers with you the day of surgery.
- DO NOT take aspirin or aspirin products within **10 days prior** to your surgery date to reduce the risk of excess bleeding.
- Do not take Motrin or other non-steroidal anti-inflammatory drugs for 10 days prior to surgery.
- Do not take Coumadin or Warfarin for 7 days prior to your surgery.

- Stop using certain herbal remedies such as Ginseng and St. John's Wort for 10 days before surgery. Additionally, vitamin E and garlic supplements should also be stopped because they can increase the risk of bleeding during surgery.

Pre-Operative Testing

Once you have scheduled your surgery, the next step is to schedule your pre-operative testing. This generally consists of chest x-ray, some blood work and an EKG (an electrical picture of the heart). Be sure to have this done one week prior to surgery so that if any of the results indicate the need for further testing, this can be done without postponing the scheduled surgery.

Full Liquid Diet:

For the 48 hours before your surgery (2 days before your surgery), please take a full liquid diet. Before you start your diet, buy 2-100oz bottles of Magnesium Citrate and a fleet enema at your pharmacy.

Permitted Foods:

Beverages:	Coffee, Tea (hot or cold), Kool-Aid, Soda, Water, Lactose-Free supplements, if recommended by your doctor, may have sugar.
Soups:	Broth, Bouillon, Fat-Free Consomme or strained cream soups.
Desserts:	Fruit Ices (without chunks of fruit), Plain Gelatin, Hard Candy, Popsicles made from juices, Custard, Ice Cream and Pudding.

Fruits/Juices:	Fruit Juices without pulp such as Apple Juice, Grape Juice, Cranberry Juice and Nectars.
Dairy Products:	Milk (whole, skim, 1% &2%), Cream.

Not Permitted

- NO alcoholic beverages 48 hours before or after surgery.
- Canned, fresh or frozen fruits, vegetables.
- Soups with vegetables, noodles, rice, meat or other chunks of food in them.
- Yogurts, breads, cereals and grain products.
- Meats, chicken, fish, and other meat substitutes (nuts, tofu, etc.).
- Oils, butter or margarine.

Itinerary

Six Weeks before Surgery

- If possible, start Kegel exercises (pelvic floor strengthening). If you are less than 6 weeks from your surgery date, start exercises now. It is not necessary to delay surgery. (See Appendix for Kegel instructions.)

Two-Four Weeks before Surgery

- Complete last round of x-rays, labs, EKG etc. per prescription. Out of town guests need to use prescriptions provided in the instructions. Call the doctor's office if you need help in any of these areas.

Two Days before Surgery:

- **Day 1: Start the clear diet**
    - Drink a bottle of *Citrate of Magnesia* at 9:00 am.
    - Drink plenty of fluids – you do not want to become dehydrated, so drink lots of fluids.

- **Day 2: Continue Clear liquids**
    - Drink a bottle of *Citrate of Magnesia* at 9:00 am.
    - Drink plenty of fluids and avoid dairy products the day prior to surgery

Day 2 Night time preparations:

- DO NOT eat or drink anything after **MIDNIGHT** the night before surgery (including water, juice, coffee, chewing gum or lifesavers)

Day of Surgery Preparation:

- Administer a Fleet's enema at 5 a.m. the day of surgery

Other Preparations:

- Leave jewelry and other valuables at home. If you wear contact lenses, glasses, or false teeth, you must bring your case to store them during surgery.
- Remove all rings from your fingers and leave them at home.

Arrival Time:

- If you are the first surgery case of the day (before 9am), then arrive at the time Ambulatory Surgery recommends.
- If you are the second or third surgery case of the day, please arrive at least 2-3 hours before surgery is scheduled. This is necessary in the event that the preceding surgeries are finished earlier than expected or cancelled for an unexpected reason.

Chapter Five
Robotic Prostate Surgery

With over a decade of experience with the da Vinci system behind them, surgeons can now approach many parts of the human body with this system. To review once again, the new da Vinci robotic system technology uses a small video camera along with customized instruments to perform minimally invasive surgery. Through small incisions, the camera and instruments are inserted, allowing a surgeon to explore the whole area to be operated on. This is accomplished without the need for a large skin incision.

When the patient has completed all the pre-operative requirements set for da Vinci robotic surgery, on the day of surgery, before arrival at the hospital (at 5 am), a fleet enema is self administered to remove any stool from the rectum. (See Appendix for fleet enema instructions.)

At the hospital, the patient will go through a check list of the pre-surgery instructions. Before entering the OR, during the 2 days before surgery, the patient will have had a bowel preparation that includes mechanical bowel cleansing. This procedure not only reduces bowel distention, but it reduces the potential for infection that could occur from the bowel in the unlikely event of bowel perforation during surgery. There will be other items to attend to, such as sequential compression stockings placed, and a large-bore intravenous line begun with preoperative antibiotics that cover the skin area.

Once properly prepared, the patient is wheeled into surgery in a hospital gown and ready for general anesthesia. The patient is usually put to sleep under anesthesia in the room after he has been transferred onto the operating room bed. Once the patient is asleep, he is positioned flat on his back with his head lower than his pelvis (this is called the trendelenberg position). His abdomen and genitalia are shaved, cleaned, sterilized and draped for surgery.

Technique of Robotic Prostatectomy - Setup

The following is the typical sequence of events during a robotic prostatectomy:

First, a small cut is made near the belly-button and then carbon dioxide is introduced into the body cavity to expand it and create a large working space. The surgeon makes two 12mm and three 8mm skin incisions (port sites) in the lower abdomen. The EndoWrist instruments and high definition camera are introduced into the body through these ports. The laparoscopic camera - a long, thin, lighted telescope - is inserted through one incision and connected to the computer monitor that allows the surgeon to see inside the body at 12-15x magnification in 3D.

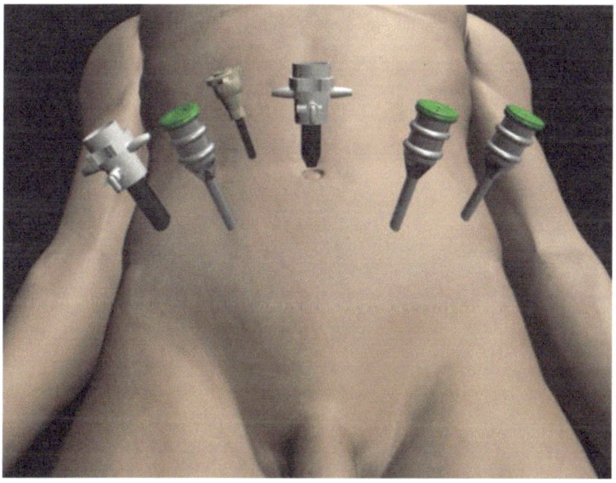

Four robotic instruments are inserted into the patient's abdomen through four laparoscopic trocars. One or two additional ports are placed for the assistant to use laparoscopic instruments.

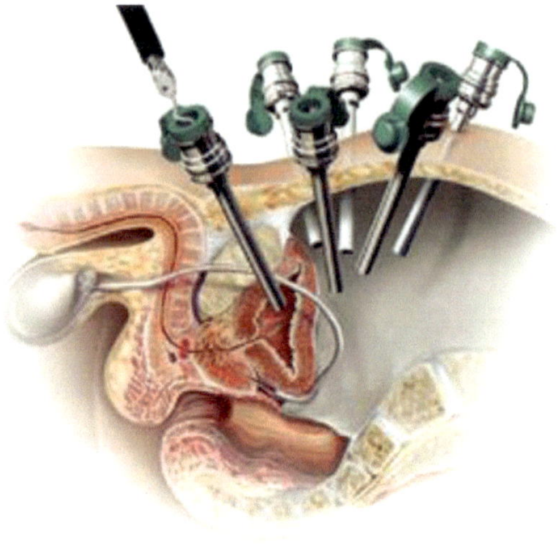

The abdomen is distended with carbon dioxide to give the surgeon space in which to work. The trocars have valves that prevent the gas from escaping. The robotic instruments are placed through the ports and manipulated inside the abdominal cavity.

Once all the ports/trocars have been placed into the abdomen, the robotic surgical manipulator is brought bedside next to the patient. The robotic arms are attached to the ports by the surgical assistant and the robotic instruments are now carefully inserted into the ports while attached to the robotic arms of the surgical manipulator.

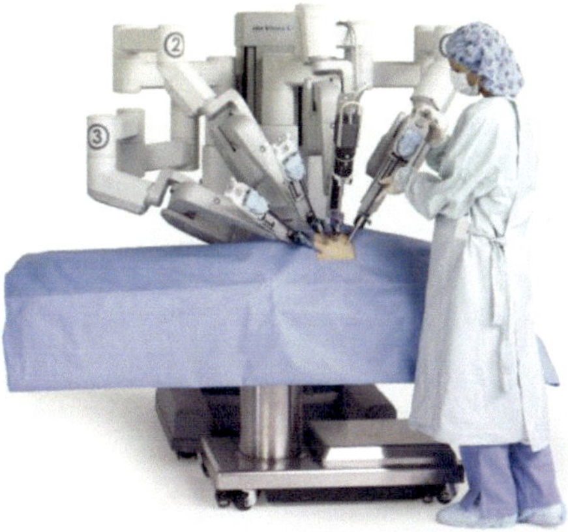

Robotic arms of the surgical manipulator are attached to the ports/trocars by the surgical assistant and the robotic instruments are now carefully inserted.

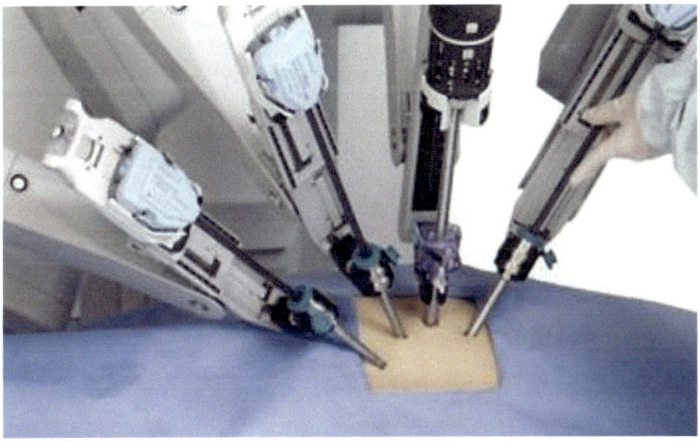

Photograph shows the connection between the robotic arm and the laparoscopic trocar. The robotic instrument enters the patient through the metal trocar.

The surgeon will now seat himself/herself at the surgical console. The surgeon will view the 3D images being transmitted by the camera inside the abdomen.

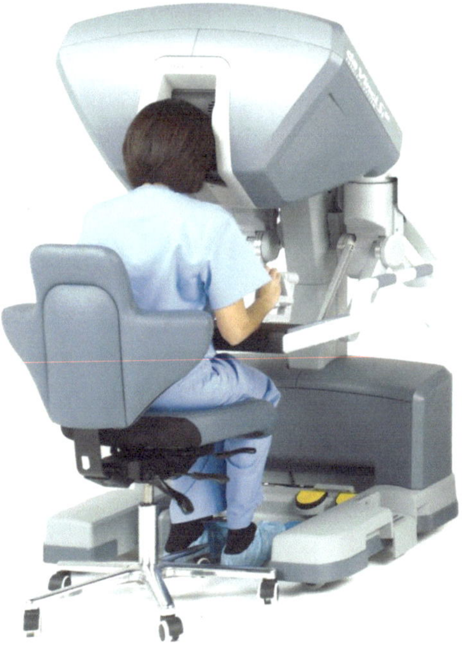

Surgeon at the da Vinci robotic system console

Once everything is in place, the surgeon will place his/her fingers in the joy-stick like hand controllers and begin surgery.

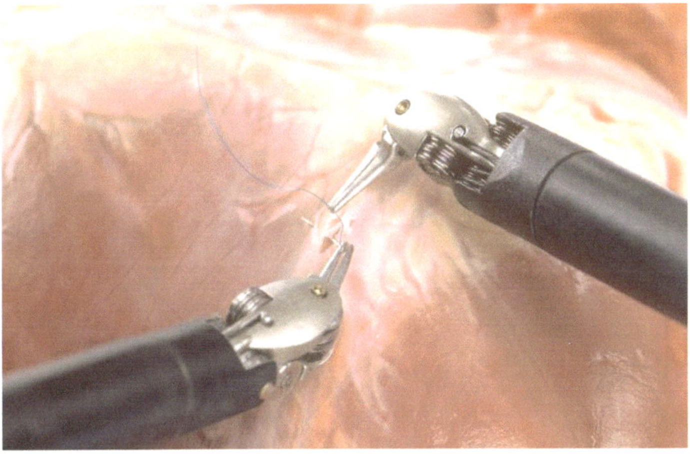

The surgeon inserts his/her hands into a "master" controller that translates motions of his/her hands into motions of the robotic arms and instruments. The surgeon acts as the "master" and the robot as the "slave" in this telerobotic "master-slave" system. The da Vinci robot only duplicates the motions of the surgeon. The da Vinci robot can not initiate any actions on its own.

Surgical Steps – Access to the Prostate

To reach the prostate, the surgeon first moves aside the bladder. He/she then works the instruments to expose the dorsal vein complex. These are large veins that drain blood from the penis and can bleed extensively during open surgery. This vein is carefully ligated to expose the prostate. The magnified vision offers careful precise suturing of this large vessel. Anterior fixation of this complex to the pubic bone can also help in promoting possible early return of urinary control for patients.

The bladder is usually separated from the prostate at the bladder neck at this point. Care is taken to preserve as much of the bladder neck as is safely possible without leaving any prostate tissue behind.

Once the bladder is separated, the vas deferens is carefully dissected free. The vas deferens is transected (this is why all men after surgery will be infertile – will not pass sperm in their ejaculate – so if the patient is planning on having his own biological children in the future, he should bank some sperm before surgery). The seminal vesicles are also carefully dissected free and are removed with the prostate. This is what leads to the lack of ejaculate when men have orgasm in the future after surgery.

Nerve-Sparing Part of Surgery

The careful lateral release of the seminal vesicles and the following lateral and posterior release of the prostate gland from surrounding tissues are very important. This is a critical point of the procedure to save erectile function. Surgeons refer to it as a nerve-sparing technique or procedure.

The cavernous nerves adhere to the lateral and posterior portions of the prostate gland and the lateral portion of the seminal vesicles as they travel to the lower pelvis and the penis. During surgery, the surgeon must carefully lift the nerve complex from the prostate gland. This delicate maneuver has been compared to peeling a wet tissue without tearing it. The cavernous nerve plexus may suffer "trauma" during surgery that requires time for recovery.

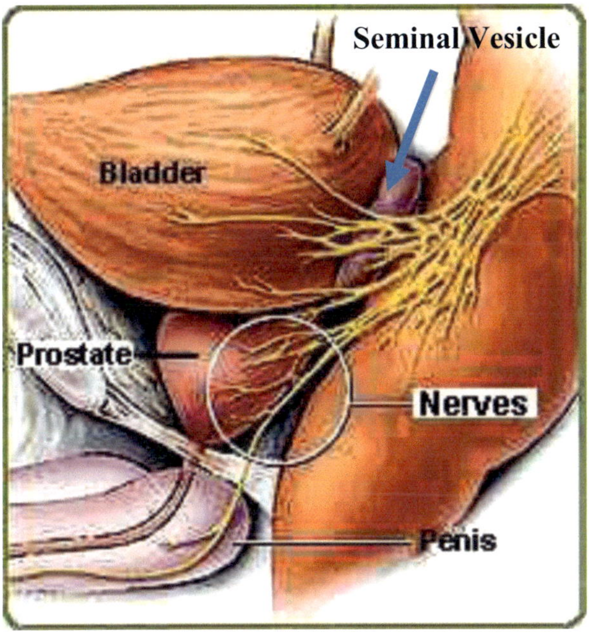

Cavernous nerve plexus on the lateral surface of the prostate gland and seminal vesicle as it heads to the penis.

With experience and skill, the surgeon can dissect very near to the prostate and carefully release the nerve plexus. This causes the nerves to drop away from the gland intact. The robot allows for stable, precise movements that only further aid the surgeon in this difficult dissection.

Prostate Removal and Urethral to Bladder Neck Reconstruction

Once the lateral and posterior attachments of the prostate have been released, the prostate is now only connected to the urethra in the lower pelvis. The urethra is carefully transected to release the prostate. Care is taken to preserve as much length of the urethra that is safely possible without leaving any prostate tissue behind. The prostate is then placed in a specimen bag and placed to the side of the pelvis (to be removed later).

The bladder neck is now reconnected to the urethra with a suture in running fashion to recreate continuity. The surgeon tries to avoid damaging any muscles surrounding the urethra, leaving the sphincter intact, and allowing early return of urinary control. The reconstruction is done over a Foley catheter, which will remain in place post-operatively.

This catheter will allow drainage of urine from the bladder out of the body without leakage at the anastomosis site. Once the anastomosis is completed, the bladder is filled with saline via the Foley catheter to make sure that there is water-tight closure. A drain is placed in the pelvis to collect any leakage of urine, if any.

The prostate is now removed in the specimen bag via the port site near the belly button (by enlarging it to 2-3cm). The prostate is usually examined by the surgeon to ensure that grossly all tumor appears to have been removed. All incisions are then closed using dissolvable sutures and skin glue. The patient is then taken to post-op recovery for 1-2 hours and then to his room in the hospital. Patients usually stay in the hospital for 1-2 days depending on how quickly they recover.

All suturing is done by robotic instruments under the direct control of the surgeon at the console.

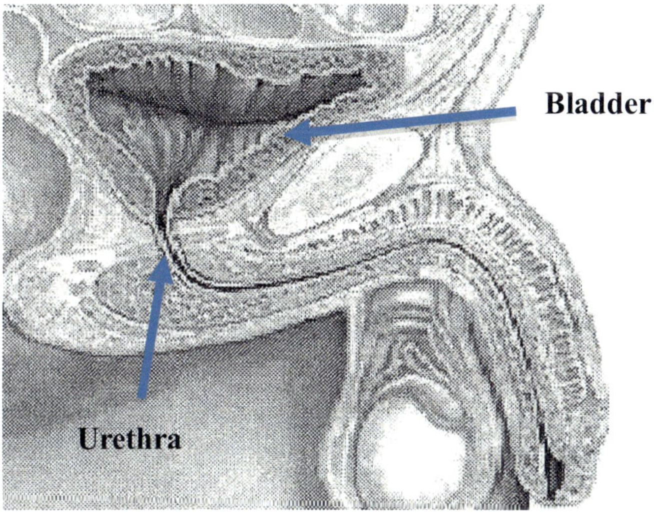

Bladder reconnected directly to urethra

The Foley Catheter

During the surgery, the surgeon will place a Foley catheter to drain the bladder as mentioned above. When the prostate is removed, the bladder has to be reconnected to the urethra as mentioned above.

This Foley catheter helps protect the connection during the healing process.

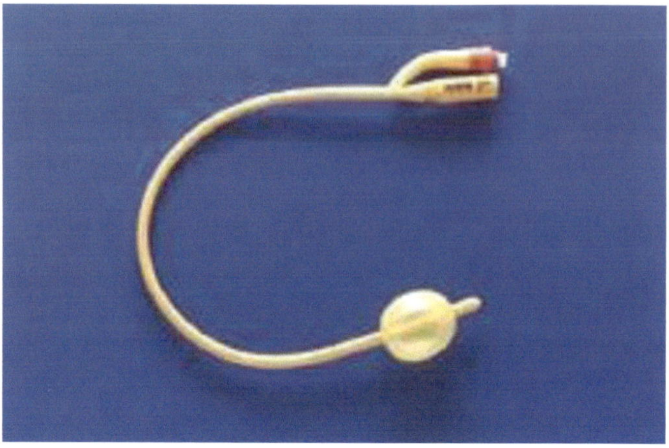

Foley catheter used during robotic prostatectomy

The Foley catheter is a soft tube that drains urine from the bladder out of the body through the urethra in the penis. In most patients, the catheter will be removed 7-8 days after surgery. Most surgeons will perform a cystogram test prior to removing the Foley catheter.

This is a test where contrast is injected via the Foley into the bladder and an x-ray picture is taken – this checks to make sure that there is no leakage of urine/contrast outside the bladder before the Foley is removed. If there is a leak, the Foley is left in place for 1 more week. Once the Foley is removed, the patient should be able to urinate freely. Sometimes, there is a small chance that it will have to be reinserted for a short while because of temporary swelling at the bladder neck connection site.

The following images illustrate how a Foley catheter works.

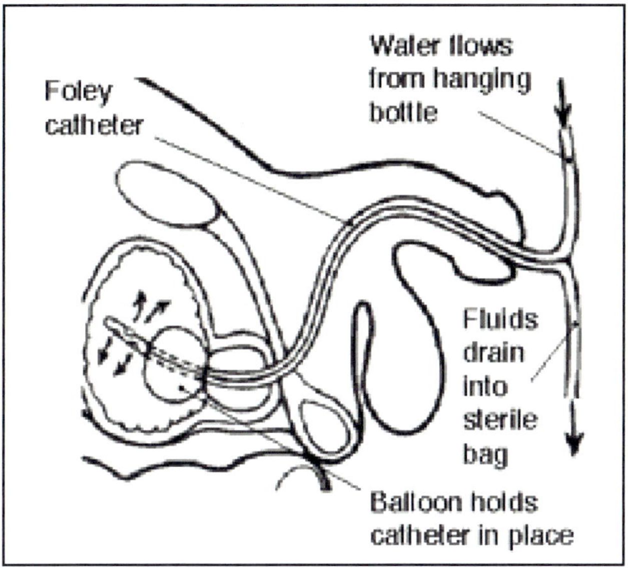

Side view diagram of male urinary tract with a Foley catheter in place to drain urine. A balloon near the tip holds the catheter in place.

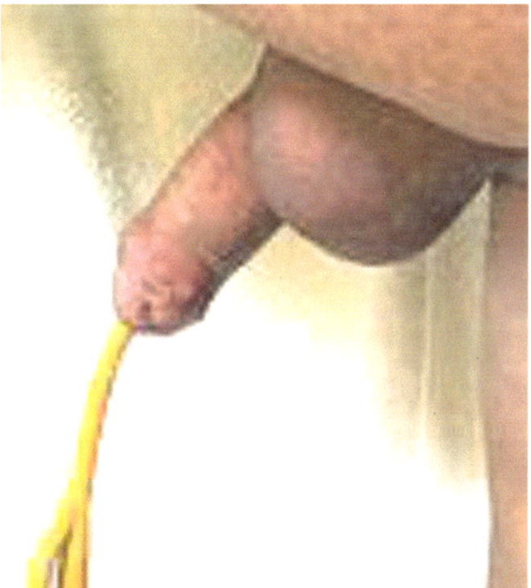

Foley catheter view outside the penis

There is a balloon on the inner end of the Foley catheter that is filled with water and prevents the catheter from falling out of the penis. There are two ports on the outside of the Foley. The larger one is connected to the drainage tube and bag to collect urine. The smaller port is what is used to keep the small inner balloon inflated. The patient should not do anything with the smaller port.

This port is only to be handled by the health care provider. The larger port connected to the drainage bag can be manipulated by the patient, for example, if changing bags or changing from the leg bag to the overnight bag and vice-versa. At home, the catheter should drain into a large bag. For outings, the patient may wear a smaller bag under the pant leg. The nursing staff will give the patient specific instructions on how to change the bags prior to discharge from the hospital.

Lymph Nodes and Their Management

The operation sometimes involves removing the lymph nodes that drain the prostate. This depends upon the specific pathology of the cancer, the PSA level and the digital rectal examination. If the lymph nodes appear normal, surgery is completed.

If the lymph nodes appear to be cancerous at the time of surgery, the surgery may be aborted depending on the extent of disease in the lymph nodes. Lymph node involvement indicates that the cancer is outside the prostate. Removing the prostate will not cure the cancer once the cancer has spread outside the prostate.

Summary of the Surgical Procedure

The operation entails removal of the seminal vesicles and prostate, while selectively preserving the muscles and nerves that help with urinary control and erectile function. After the prostate is removed, the bladder is re-attached to the urethra with stitches.

Because of the removal of the seminal vesicles and the transection of the vas deferens, men can no longer naturally father children following the procedure. However, with new nerve sparing techniques, many men will regain erections and orgasm function after surgery.

Patients usually are discharged in 24-48 hours and return to normal activity in one to two weeks. They usually return to strenuous activities in three to four weeks. There are variations and complex steps to the surgery, but the basics are listed here in this chapter. The average hospital stay is brief and uneventful. When the patient awakens from the anesthesia, he will feel a little soreness. The catheter will be in the penis to carry urine to a bag. The catheter needs to be in place until the urethra heals, usually 7-8 days. Within a few hours of the surgery, most patients are able to move around and have a liquid diet.

Chapter Six
Post-Operative Instructions

Because da Vinci Robotic Surgery is performed through very small incisions, it is usually associated with minimum surgical pain. Most patients recover with minimal narcotic medication, which reduces side effects such as dizziness, lethargy, and constipation. The reduction of pain also permits most patients to get on their feet within hours following surgery and to leave the hospital on the first post-operative day.

However, surgery is never a walk through the park and for most patients robotic prostatectomy is the first major surgery they are ever having in their lives. This chapter goes over some of the post-operative issues that may arise to give the patient a realistic expectation of what may occur.

Bladder Spasms

Bladder spasms are commonly experienced as a moderate cramping sensation in the lower abdomen or bladder and are quite common after prostatectomy. These spasms are usually transient and often decrease over time. If severe, medications can be prescribed by your doctor to decrease the episodes of these spasms. The patient can expect to have a urinary catheter (Foley) draining your bladder for approximately 7-8 days after the surgery.

Blood in the Foley Catheter and Drainage Bag

It is not uncommon to have blood-tinged urine for a few days to a week after your surgery. If the patient starts to pass large clots or the Foley catheter stops draining (the patient feels like he has to urinate but no urine is draining through the Foley), then he should go to the emergency room or contact your physician immediately.

In no circumstance is the patient to let anyone take the Foley catheter out except with the direct permission of the surgeon who did their surgery. This is because, if the catheter is taken out prematurely before the bladder to urethra connection has healed, this may cause scarring and permanent loss of urinary control in the future for the patient.

Urinary Catheter (also called a Foley catheter)

Putting antibiotic ointment (i.e., Neosporin) on the tip of the penis a few times a day while the catheter is in place can help to reduce the discomfort at the tip of the penis that some patients experience. It is normal to have urine leak around the catheter as well as through it, so be sure to line underwear with pads even with the catheter in place.

It is also normal, especially as you become more active, to have some blood in the urine. If there is blood in the urine, be sure to hydrate yourself well to flush any clots out. Again, as mentioned above, in no circumstance is the patient to let anyone take the Foley catheter out except with the direct permission of the surgeon who did their surgery.

At home, you'll need to take good care of your catheter. Examine it regularly to make sure it is securely in place. It may be wise to clean the opening of the penis with a 50/50 solution of water and hydrogen peroxide or an antibiotic ointment that your doctor may recommend. The catheter will drain into a large bag. When you want to go out, you can wear a smaller bag under your pant leg.

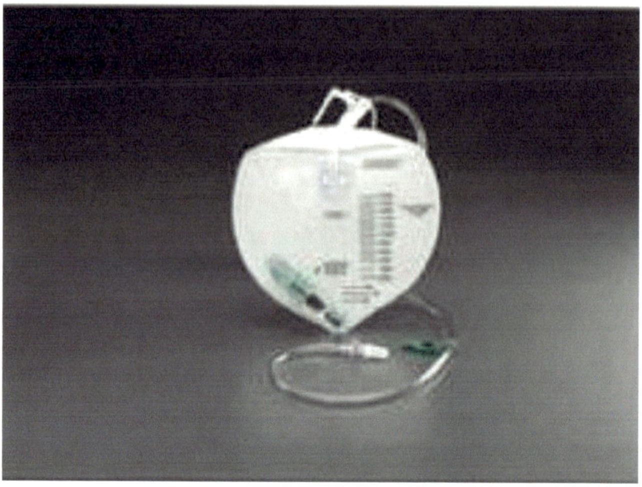

Large over-night urinary drainage bag

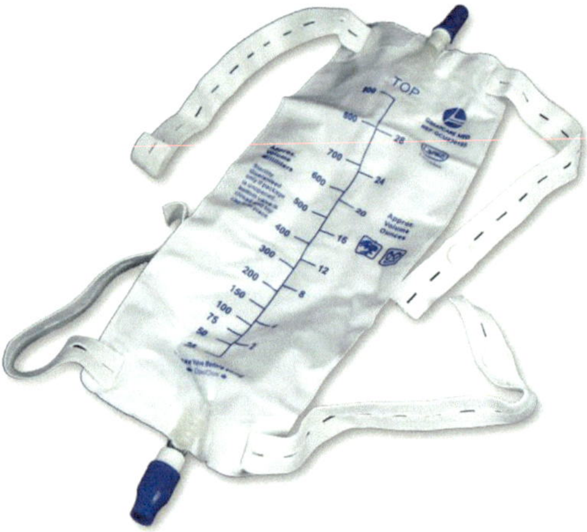

Leg urinary bag when patient is out and about (ambulatory)

Pelvic or JP (Jackson-Pratt) Drain

The pelvic drain is placed in the operating room and drains the pelvic space around the bladder-urethra anastomosis. This drain is usually removed in 24 hours when the drainage is minimal.

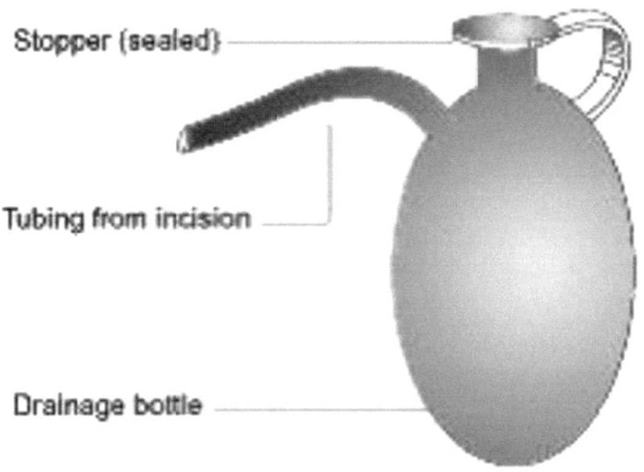

JP Drain Bulb

Diet

Eat clear liquids such as jello, broth, or juices (no soda or carbonated beverages) until you are regularly passing gas without difficulty or until you have a good bowel movement (this may take a few days). You may then resume a regular diet. Avoiding gas-producing foods, such as beans and broccoli, is still recommended.

Activity Level

- It is good for you to walk around.
- Do not sit in one place for long periods of time.
- Absolutely no biking, motorcycling, or horseback riding for 4 weeks.
- You can do as much walking as you can tolerate.

- Do not climb stairs excessively for 2-3 weeks after surgery.
- You may take a shower 24 hours after surgery. No tub baths, swimming or hot tubs for 2-3 weeks after surgery.
- Do not drive while taking pain medications.

Skin Integrity

- You will have 5 port sites (small incisions that the doctor performs the surgery through) that will either have skin glue or have steri-strips (small pieces of tape) and Band-Aids over them.
- Band-Aids may come off in 48 hours. Steri-strips may stay in place until you are seen in clinic (1-2 weeks after surgery).
- If skin glue is used – leave the incisions alone for 1-2 weeks. The glue will fall off on its own after 1-2 weeks. Do not use any ointments or alcohol to clean the incisions – this will erode the glue and cause it to fall off too early.
- Once your dressings are off, it is not uncommon to have a very small amount of drainage from where your dressings were. There are no staples or stitches to be taken out. All stitches are self-dissolving.
- It is normal to have some bruising of the skin on your abdominal wall. This can develop anywhere from 1-5 days out from surgery. Although it can look scary, it is rarely anything to worry about and resolves in a few weeks.

Clothing

After surgery, your abdomen may be bloated, making it difficult to fit into regular button pants. It is recommended that recovering patients wear pants with elastic waist for comfort. Wear loose, comfortable clothing.

Medications

- It is possible to resume daily medications as soon as you are discharged from the hospital. The only exception to this is Coumadin which is generally resumed after the Foley catheter is removed.
- An antibiotic will be prescribed, to be taken by mouth. Start this the day you leave the hospital and continue taking this medication until 3 days after the catheter is removed. You will receive a prescription the day of your discharge.
- An anti-inflammatory pain medication will be prescribed for you, to be taken by mouth daily and an additional stronger pain medication will be prescribed to use if needed. You will receive these prescriptions the day of your discharge.
- A medication to reduce the frequency and severity of bladder spasms will also be prescribed and should be taken every morning until the catheter is removed.
- A stool softener should be taken by mouth two times daily. Senokot-S or Colace are good choices. You can buy this medication over the counter and do not need a prescription. All narcotic pain medications are constipating and a stool softener will help to prevent this.

Recovery Time

Recovery time can vary, depending on the health of the patient prior to surgery. The following is a list of what to expect:

- Most surgeons complete surgeries in a few hours. Hospitalization usually lasts less than 24 hours. All patients go home with a catheter in place continually draining the urine into a drainage bag.

- You will be seen about 8 days after surgery at The Urology Team Office to determine if the catheter will be removed at this visit. An x-ray may be done that morning of your appointment, at a separate location, to help in deciding if the catheter can be removed.
- You can return to light activity within 24 hours and will be cleared by the surgical team to return to unlimited activity generally 3-4 weeks after the surgery. Most men have difficulty with urinary control at the beginning and will require some form of protection, such as a pad that fits inside your underwear. That is why it is important at the first visit to bring Depend Guards for Men or urinary pads and a couple of pairs of Jockey underwear.
- Within a few weeks to months, you should be seeing improvement in your urinary control. Sometimes, the recovery of continence is slower, so be patient and continue your physical therapy and exercises.
- Kegel exercises should be started/resumed after the Foley catheter is removed. These exercises help to regain your continence. At first, it may be hard to find these muscles, but this can be done by starting and stopping your urine stream. Once you find the correct muscles, repeat the flexing and relaxing of these muscles without urinating. Begin by squeezing the muscles for a count of 3, and then relax for a count of 3. Work up to repeating these exercises for 2-3 minutes two to three times a day. These will help to strengthen your muscles around the bladder that help hold the urine. For more detailed information on how to perform Kegels, please refer to the Appendix.

- The recovery of potency after a prostatectomy can be slow and time-dependent. Even though the nerves to the penis can be spared, there is still some injury from trauma or stretching from the operation. These damaged nerves need time to heal. At each follow-up visit, issues regarding sexual function will be discussed with you by your health care provider.

Decrease in Penis Size

In the *Journal of Urology*, it is noted that some men have reported a relative decrease in penis size following prostatectomy for prostate cancer. This may be due to the way the bladder is reconnected to the urethra or due to penile atrophy. There is very little research on this topic, but your surgeon will go over strategies to minimize this.

Things You May Encounter after Surgery

- **Bruising around the Incision Sites:** Not uncommon and should not alarm you. This can look scary but is quite normal and will resolve over time.
- **Abdominal Distention, Constipation or Bloating:** Make sure you are taking your stool softener as directed. If you don't have a bowel movement 48 hours after surgery, try taking Milk of Magnesia as directed on the bottle.
- **Weight Gain:** Do not be alarmed. This is temporary due to the gas and fluid shifts. Your weight will be back to your pre-operative weight in generally 5-7 days.
- **Scrotal/Penile Swelling and Bruising:** This is not abnormal and should not alarm you. It may appear immediately after surgery or may start 4-5 days after surgery. It should

resolve in about 7-14 days. You may also try elevating your scrotum on a small towel or washcloth that you have rolled up when you are sitting or lying down to decrease the swelling. It is also recommended to wear Jockey or snug-fitting underwear for support, even with the catheter in place.

- **Bloody drainage around the Foley catheter or in the urine:** Especially after increasing activity or following a bowel movement, this is not uncommon. While this is often alarming, it is not uncommon and usually resting for a short period of time improves the situation. Call your surgeon if you see lots of clots in your urine or if you have no urine output for one to two hours.
- **Bladder Spasms:** It is not uncommon with the catheter in and even after the catheter comes out to have bladder spasms. You may feel mild to severe bladder pain or cramping, the sudden, urgent need to urinate, or a burning sensation when you urinate. You will be prescribed a medication called Detrol LA to take once a day to help reduce this discomfort.
- **Perineal Pain** (pain between your rectum and scrotum): Perineal discomfort may last for several weeks after surgery, but it will resolve. Call your surgeon if the pain medication does not alleviate this. You can also try elevating your feet on a small stool when you have a bowel movement, using Anusol ointment, and increasing the fiber and water intake in your diet. You may also benefit from using a donut shaped pillow to sit on.
- **Lower leg/ankle swelling:** This is not abnormal and should not alarm you. It should resolve in about 7-14 days. Elevating your legs while sitting will help. If you develop pain in the back of your leg (Charley-horse) – please inform your surgeon immediately or go to the emergency room. This

may indicate a clot in your lower leg and should be treated immediately.

Urine Leakage During Intercourse

There is no ejaculate during intercourse, but urine leakage may occur during intercourse. This may disturb the man more than the woman. Urine is harmless, just inconvenient.

Follow-Up Appointments

- **7-8 days after surgery:** A cystogram (an x-ray of your bladder) will be taken at an x-ray facility to see if enough healing has occurred in order to remove the Foley catheter. Later that day, you will have a follow-up visit with the doctor. At this appointment, you will need to bring a pair of Jockey underwear and several Depend Guards for Men or urinary pads. Remember to continue your antibiotics until the catheter is removed and then for 3 days following the catheter removal.
- **2-3 months after surgery and every six months after that:** You will continue to follow up with your doctor and will get a PSA drawn prior to each appointment to look for evidence of recurrence or re-growth of the tumor.

Contact your surgeon immediately or go to the emergency room if you experience any of the following symptoms:

- Temperature over 101 degrees F

- Urine stops draining from your catheter into the drainage bag
- Any pain so excruciating that pain medication is not relieving it
- Large amount of blood clots in urine.
- Bladder spasms that are not relieved with pain medication
- Pain in the back of your leg (Charley-horse)
- Difficulty breathing or chest pain

Pain Management

Consult with your doctor about pain management. Most patients do well with over-the-counter medications like Ibuprofen, others need stronger prescription medications, usually immediately following surgery.

Avoid Heavy Lifting

Don't think you are in shape for heavy work just because you feel great after surgery. You will need to rest and allow your surgery to heal. Don't lift anything heavier than 10 to 15 pounds for at least four to six weeks. If your job does not involve heavy lifting or fast movements, then it is okay to go back to work fairly soon after surgery.

Keep in mind that your body is a marvelous instrument that heals and returns to normal given a little time to recover from surgery. Don't expect to have a rebounding of your usual self in the first several weeks after surgery. It takes time to heal.

Chapter Seven
Prostate Surgery Rehabilitation

Sex after Surgery

For men facing prostate cancer, their first worry is about curing the cancer. With many, their second worry is whether the treatment will destroy their sex life. After prostate cancer surgery, a normal sex life depends on a number of variables. A man's sex life before surgery, whether he has a supportive partner, the stage of his cancer, and even how a surgeon defines potency all will influence the odds he is given by his doctor.

It is important to remember that regaining erectile function takes time after prostate surgery. Nerve tissue, even when spared, takes time to recover from the inflammation and scarring that can occur after surgery. This is going to take time to heal. You should try to remain patient, and remember that this will be an ongoing rehabilitation program that will last for 6-12 months. Rehabilitation of potency is much like a sports injury. Proper conditioning and medical treatment may potentially speed up recovery, but it is not guaranteed.

Viagra

Viagra 100mg will be prescribed and should be taken every other evening on an empty stomach just before bedtime. The reason for this is that during your sleep cycle, your brain stimulates nighttime erections. Immediately after surgery, it is unlikely that you will have rigid nighttime erections, however, taking Viagra at night allows for improved blood flow to the penis during these stimulation periods.

This improved blood flow has been shown to have some significant benefits to the recovery of your sexual function. Specifically, following this regimen for 6 months has been proven to reduce scarring in your erectile tissues and retain the smooth muscle necessary for erections to occur. Four weeks after surgery, you will have healed up enough to attempt sexual stimulation. Although some patients are able to achieve erections satisfactory for sexual activity and climax this early after surgery, just seeing "some action" or activity in the penis at this point is your goal and is expected to improve over the next 6-12 months. The use of Viagra is the “active” part of your rehabilitation regimen because it involves your body actively pumping good oxygenated blood to your penis.

Most insurance companies only allow for 6-8 pills of Viagra a month. However, this regimen requires 15 pills per month. This additional cost for the Viagra is usually paid out of pocket by the patient.

Vacuum Pump Erection Devices

The “passive” portion of your erectile rehabilitation program involves the use of vacuum pump erection devices. At 4 weeks, you will begin using these devices twice a day to passively pump good oxygenated blood into your erectile tissues and holding it there for 5 minutes. These devices use a vacuum mechanism to pull healthy blood into the erectile tissue of your penis, helping to stretch out these tissues and keep them healthy. Most programs will provide you with an instructional DVD on the use of these devices and you will be scheduled to meet with the vacuum pump company Clinical Specialist. This specialist will act as your "personal trainer" for this portion of your erection rehabilitation program. The purpose of this consultation is to focus on the correct use of the vacuum pump device and maximize the results of your erectile rehabilitation program. Below is an image of one commercially available vacuum pump device.

Osbon Erecaid

How Will All This Help Me?

Recent clinical studies have demonstrated that penile rehabilitation through the use of Viagra and exercise programs following prostate surgery can significantly enhance the ability of the patients to regain their pre-surgery sexual function.

If you don't use it, you lose it! After prostate cancer treatment, many men see a decrease in penile length and girth. With this program, you are more likely to avoid these side effects. It's important to maintain a healthy penis during this phase in order for the penis to function properly in the long run. Playing an active role in your recovery process is critically important to maximizing your chances of achieving the best outcome possible.

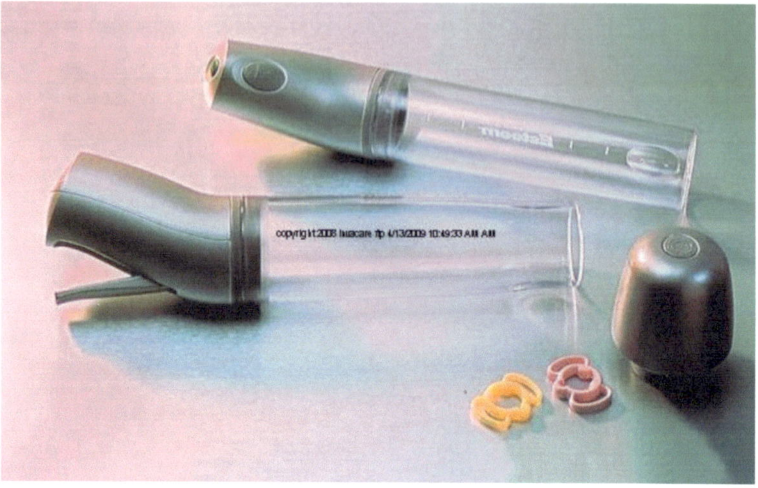

Osbon Erecaid variety of hand-held vacuum devices

Using this vacuum device requires some assembly; essentially, a cylinder and a pump are attached together, with a penile elastic tension ring that can be adjusted. The penis is inserted into the cylinder in such a way that it is possible to maintain an airtight seal; then the pumping begins until erection occurs.

During the rehabilitation program, the penile elastic ring is not used. However, when attempting to use the vacuum pump to achieve an erection for intercourse, a penile elastic ring can be placed at the base of the penis after the erection is achieved to help maintain the erection. This ring should only be used occasionally for intercourse, and not twice a day for the rehabilitation program – the rehab program is designed to simply engorge the penis twice a day with oxygenated blood to nourish the penile tissues.

Chapter Eight
Frequently Asked Questions

Q: How many procedures has the surgeon performed?
A: Frankly ask the surgeon during consultation the number of da Vinci robotic prostatectomies that he/she has individually performed. It is usually recommended that the surgeon perform 250 procedures to become highly skilled.

Q: Does the da Vinci robotic approach result in a higher cancer cure rate than conventional surgery?
A: Because of the precision of the technique and the magnification of the da Vinci laparoscopic camera, the completeness of cancer removal is reported in the literature to be as good as open surgery.

Q: How long does the operation take?
A: Barring unusual circumstances, most surgeons handle the total procedure in a few hours (2-3 or less).

Q: Will I need to donate my own blood to have on hand during surgery if the need for a transfusion arises?
A: One of the many benefits of the da Vinci robotic approach is minimal loss of blood. It is rare to need a blood transfusion. Therefore, it is usually not necessary to donate your blood.

Q: Will this operation render me sterile?

A: Yes, all patients undergoing radical prostatectomy will not be able to father children after the procedure. However, the nerves that give the ability to have erections can be spared, allowing many patients to continue to have erections and orgasm after the surgery. The only way to father children after surgery would be to retrieve sperm directly from the testicle and then use this sperm for in-vitro fertilization (IVF) with your female partner which can be quite expensive. If you are thinking about having children in the future, you may bank sperm before the surgery for IVF at a later date.

Q: How long can I expect to stay in the hospital after the procedure?

A: Most da Vinci robotic prostatectomy patients have been able to go home the day after surgery. Sometimes, patients may stay in the hospital for 1-2 days. What is most important is your comfort and safety and you will only be discharged when you are ready.

Q: After the da Vinci robotic prostate surgery, when will I be able to resume normal activities?

A: One of the advantages of the da Vinci robotic approach is decreased pain which results in easier recuperation. Thus, most individuals undergoing this operation will be able to resume light activities within 24 hours of surgery and can resume unrestricted activity in 3-4 weeks. However, each individual heals at his own rate. You should listen to your body, and do what it tells you to do.

Q: After the da Vinci robotic prostate surgery, will I be able to have normal sexual relations immediately after surgery?

A: No. Even when the nerves are preserved, it takes time for them to heal and function. With the extended nerve sparing technique, many patients are able to return to sexual activity, but this recovery varies based on the patient's age, whether one or both nerves were spared, and what their pre-surgical level of erectile function is.

Q: How long can I expect to be off work after the procedure?
A: Many patients (including physicians, surgeons, and lawyers) are able to return to light work in one week. However, many patients take time off for one or two weeks after the da Vinci robotic prostate surgery. Long-term disability is not necessary after this procedure, since complications are rare and recovery is typically rapid. Any further time off will be recommended if medically necessary.

Q: How safe is the da Vinci robot for laparoscopic prostatectomy?
A: The da Vinci is FDA approved for radical prostatectomy and is being used routinely in over 300 locations world wide. (See intuitivesurgical.com for a current listing). Each da Vinci system is rigorously maintained, tested, and upgraded as necessary by Intuitive Surgical. Equipment failure is exceedingly rare.

Q: What happens if there is a malfunction in the da Vinci system?
A: In the unlikely event of malfunction, or if the surgeon feels that it is not safe to continue with the robot, the da Vinci system will be withdrawn and the surgery can proceed laparoscopically without it. The instruments and supplies necessary to perform the surgery without the da Vinci are kept on hand so that conversion, if necessary, can occur seamlessly.

Q: My doctor tells me that the surgeon loses tactile sensation with robotic surgery. How does this affect the outcomes of surgery?

A: Tactile sensation, the ability to perceive through touch, is an important part of open radical prostatectomy surgery. The surgeon usually uses this sense to palpate the edges of the prostate to assess if there is any cancer. However, most robotic prostatectomy patients have organ confined disease, and this tactile feedback of the edge of the prostate is not as critical. With da Vinci robotic surgery, true tactile sensation is lost. But the ability to move the robotic camera to difficult locations within millimeters of where the surgeon is working allows him/her to see things an open surgeon cannot, and do so with the 12-15 fold magnification of the robotic camera. With experience, a pseudo-tactile sensation is developed and an appreciation of tissue texture and density is achieved. The combination of enhanced vision and experience with tissue handling compensates for the lack of real tactile sensation. This is why the cancer results for da Vinci laparoscopic robotic surgery are just as good as open surgery.

Q: Many urologists feel the benefits of da Vinci robotic surgery are unproven. What is your opinion?

A: Open radical prostatectomy is a highly sophisticated and finely developed operation. When it comes to surgery, whether you perform it through an open incision or via the da Vinci robotic technique, the same things matter: operative time, blood loss, negative margin rates, post-surgical pain, time to recovery of urinary control, time to recovery of erections, time until recovery to full physical activity. All of this information has been published and is available on da Vinci robotic surgery and it shows equal and, in some cases, better outcomes when compared to open surgery. The beauty of da Vinci robotic surgery is that these outcomes are accomplished while minimizing bleeding and pain.

Q: My urologist tells me that it takes 15 years to know if a prostate cancer treatment really works. How can you tell that the da Vinci surgery works if it's only been around for 10 years?

A: Waiting 15 years to analyze prostate cancer cure holds true for treatments that do not remove the prostate, such as radiation, because you need to wait to see if all the cancer was killed or if it will eventually return in the prostate. With surgical removal of the prostate (da Vinci or the old open technique), the prostate is removed and once it's out, you can find out the surgical margins.

Surgical margins are the edges of the prostate gland, and the pathologist will look at this when the prostate is removed. If your surgical margins show no cancer at the edges, there is a very high likelihood that the surgeon got it all out locally. This does not mean it has not spread to other parts of the body. It just means there is nothing left behind in the spot where the prostate used to be. We have data from thousands of laparoscopic and da Vinci robotic prostate surgeries to show that the surgical margins are equal to, and, in some series, better than, what we were seeing with the old open surgery. If the surgical margins are the same as open surgery, then the only way the cancer can come back is if it spread before the surgery was done. This pre-surgery spread of disease will be the same no matter how the surgery is performed.

Glossary of Terms

3D (three-dimensional) vision: Stereo or binocular vision created by two optical channels, as in two eyes. Three-dimensional vision blends the separate images seen by each eye into one composite image, which gives us the ability to perceive depth and judge distances.

Ablative therapy: Treatment that destroys the function of an organ without removal. For example, high intensity focused ultrasound (HIFU) ablates the prostate – it destroys the cancer, but the prostate is not removed, it still lingers as a scarred, non functional organ.

Articulating: A movable, rotating joint, as in the human wrist or as in *da Vinci*'s patented *EndoWrist*® instruments

Benign: Referring to abnormal cell growth that is not cancerous, i.e., not dangerous to health, not recurrent or progressive (especially of a tumor).

Biopsy: A procedure in which a sample of tissue is removed and then examined under a microscope. The doctor will use transectal ultrasound (TRUS) for guidance and insert a narrow needle through the wall of your rectum into several areas of your prostate gland. The needle then removes a cylinder of tissue, usually about 1/2-inch long and 1/16-inch across, that is sent to the laboratory to see if cancer is present.

Bladder cancer: Cancer of the bladder, the organ that stores urine until it is passed out of the body.

BMI: Body Mass Index (BMI) is a number derived by using height and weight measurements that gives a general indication if weight falls within a healthy range.

Cancer grading: If your biopsy shows the presence of prostate cancer, the pathologist assigns each tissue sample a grade, indicating how far the cells have traveled along the path from normal to abnormal.

Cancer staging: The process used to find out if cancer has spread within the prostate or to other parts of the body.

Cardiac: Referring to the heart.

Cardiothoracic conditions: Conditions affecting the heart and surrounding structures located inside the chest cavity or ribcage (thorax).

Chemotherapy: Chemotherapy, which kills fast-growing cells, has not proven particularly effective against slow-growing prostate cancer cells. Several promising new anti-cancer drugs are under study, being added to either surgery or radiation therapy for men with Stage III prostate cancer. Chemotherapy is also being tried in conjunction with hormonal therapy for men whose advanced cancers are no longer responsive to hormonal therapy alone.

Conformal radiation therapy: A 3-dimensional conformal radiation therapy (3D-CRT) uses sophisticated computer software to conform or shape the distribution of radiation beams to the 3-dimensional shape of the diseased prostate, sparing damage to normal tissue in the vicinity of treatment.

Conservative management: Also called watchful waiting, conservative management, observation or surveillance, is based on the premise that cases of localized prostate cancers may advance so slowly that they are unlikely to cause men—especially older men—any problems during their lifetimes. Men who opt for conservative management have regular medical checkups and are asked to report any new symptoms to the doctor immediately.

Cryosurgery: A procedure that uses liquid gases to freeze and kill prostate cancer cells. During cryosurgery, a warming catheter inserted through the penis protects the urethra, and incontinence is seldom a problem. However, the overlying nerve bundles usually freeze, so most men become impotent. The procedure takes about 2 hours and requires anesthesia (either general or spinal).

Digital rectal exam (DRE): During this exam, a physician inserts a gloved, lubricated finger into the rectum to feel for any irregularity that might suggest the presence of prostate cancer. The prostate gland is located just in front of the rectum, and most cancers begin in the back part of the gland that can be reached by a rectal exam. Although DRE is less effective than the PSA blood test in finding prostate cancer, it can sometimes find cancers in men with normal PSA levels.

Endoscope: A small tubular lens device, usually attached to a video monitor, and video camera, and used to visualize the inside of the body.

Hormonal therapy: Hormonal therapy combats prostate cancer by cutting off the supply of male hormones (androgens) such as testosterone that encourage prostate cancer growth. Hormonal control can be achieved by surgery to remove the testicles (the main source of testosterone) or by drugs.

Indication: A reason to prescribe a drug or perform a procedure; "the presence of bacterial infection was an indication for the use of antibiotics".

Kidney: The organs that remove waste from the blood, contribute to red blood cell production and blood pressure regulation.

Laparoscopy: A type of surgery using a surgical camera called a laparoscope or endoscope, which is inserted into one or more small incisions, to see inside the abdominal cavity. The surgeon views the affected area on a standard 2D video monitor.

Laparotomy: Open abdominal surgery using a large incision.

Lymph nodes: Bean-shaped structures scattered along vessels of the lymphatic system. These nodes act as filters, collecting bacteria or cancer cells that may travel through the body.

Malignant: Characterized by progressive and uncontrolled cell growth (especially of a tumor); cancerous.

Minimally invasive surgery: Minimally invasive surgery (also known as MIS, minimal access, keyhole surgery, laparoscopic, or endoscopic surgery) is intended to eliminate the most physically and emotionally traumatic elements of conventional open surgery, including pain and lengthy recovery. Advances in MIS allow surgeons access to affected tissues and organs through very small incisions.

Obesity: Having a body mass index (BMI) greater than 40, or a BMI of 35 with two or more related health problems that increase risk for early death.

Open surgery: Any surgical procedure performed using a large incision made for direct access to the affected organ or tissue.

Port: A tiny (1-2 cm) incision, into which a cannula (hollow, rigid tube) is inserted to act as a conduit for an endoscope or minimally invasive surgical instruments, including *da Vinci*'s *EndoWrist*® instruments.

Prostate: The prostate gland, a key part of the male reproductive system, is linked closely with the urinary system. It is a small gland that secretes much of the liquid portion of semen, the milky fluid that transports sperm through the penis during ejaculation. The prostate is located just beneath the bladder, where urine is stored, and in front of the rectum. It encircles a section of the urethra.

Prostate cancer: Cancer of the prostate, the male reproductive gland located below the bladder & in front of the rectum. Early prostate cancer usually causes no symptoms and is found by a PSA test and/or DRE.

Prostatectomy: Surgery to remove the prostate in order to treat and prevent the reoccurrence of prostate cancer.

Prostate-specific Antigen (PSA): Testing a blood sample for the amount of PSA plays an important part in the early detection of prostate cancer. Prostate-specific antigen (PSA) is a substance made by the normal prostate gland and a small amount is found in blood. Most men have levels under 4 nanograms per milliliter (ng/mL) of blood. When prostate cancer develops, the PSA level usually goes above 4.

Radiation therapy: Radiation therapy uses high-energy x-rays, either beamed from a machine or emitted by radioactive seeds implanted in the prostate, to kill cancer cells. When prostate cancer is localized, radiation therapy serves as an alternative to surgery.

Robot-assisted or robotic surgery: Surgery performed using the assistance of robotic technologies. Existing systems like *da Vinci* are designed to provide surgeons with enhanced capabilities rather than designed to operate on their own.

Ureter: One of a pair of tubes, about one foot long, that carries urine from each kidney to the bladder.

Urethra: The tube that carries urine from the bladder out through the penis. During ejaculation, semen is secreted by the prostate through small pores of the urethra's walls.

Additional Information & Appendix

What Is the Prostate Gland?

The prostate is a gland—about the size of a walnut—that makes and stores semen. The prostate is necessary for male sexual function and reproduction.

What Does It Do?

The main job of the prostate gland is to produce semen, the fluid that carries sperm. This fluid helps protect sperm from the vagina's acidity during ejaculation.

Where Is the Prostate Gland Located?

The prostate gland is located just below the bottom of the bladder and in front of the rectum. It surrounds the urethra, which is the tube that carries urine from the body. Because of its location, if the prostate gland becomes enlarged, it can create bothersome urinary problems.

Anatomy of the Prostate Gland

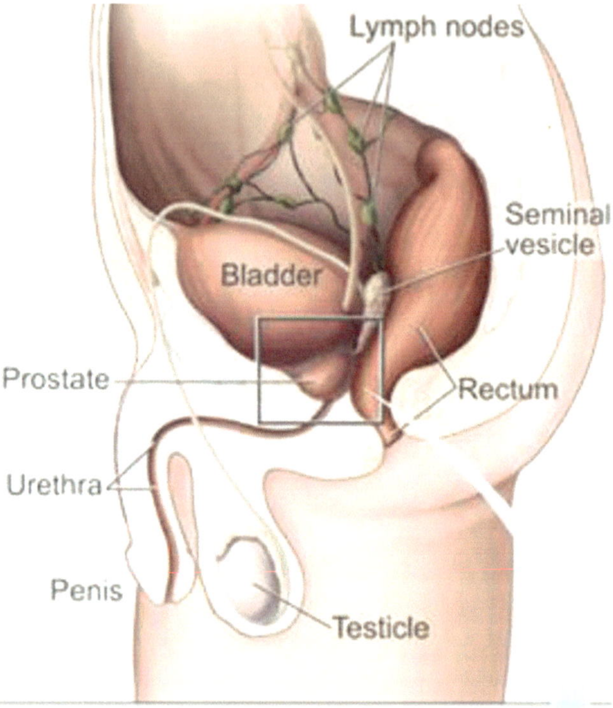

This shows the prostate and nearby organs.

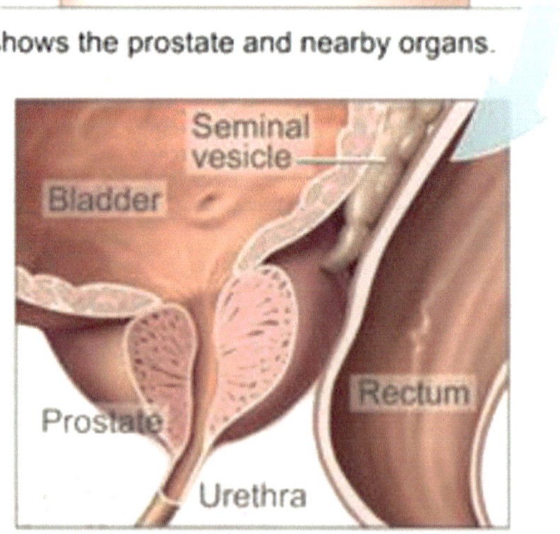

This shows the inside of the prostate, urethra, rectum, and bladder.

Grading of Prostate Cancer

What is the Gleason System for grading cancer?

One way of grading prostate cancer is the Gleason System. This grading system is based on a number range from 1 to 5. The lower the number, the lower the grade, and the slower the cancer is growing. The higher the score, the higher the grade of the tumor. High-grade tumors grow more quickly than low-grade tumors, and are more likely to spread to other parts of the body. The Gleason score is reported as two numbers, for example: Gleason 3+3=6. The two score numbers represent the major and minor components of the cancer on the biopsy reviewed by the pathologist. So for example, if the biopsy shows a lot of Gleason 4 and just a little Gleason 3, the total Gleason score is reported as 4+3=7. Thus, a Gleason score of 3+4=7 is better than a Gleason score of 4+3=7, since the latter has more Gleason 4 component.

- Grades under and equal to 6 mean that the cancer is likely to be less aggressive.
- Grade 7 is in the intermediate range. This means that the cancer cells are more likely to be aggressive and grow faster.
- Grades 8 to 10 indicate that the cancer cells are more likely to be very aggressive in growth.

Courtesy of Ohio State University

Statistics on Prostate Cancer

Although researchers found no consistent evidence of a relationship between prostate cancer risk and either body mass or physical activity, increased risk of prostate cancer was found to be associated with high intake of saturated fat in each of the ethnic groups studied. Other factors, such as genetically determined hormone levels and diet during adolescence, may account for differences in incidence among the ethnic groups studied.

Five-year survival rates are lower for African-American men (66.4 percent during 1983 to 1990) than for white men (81.3 percent during 1983 to 1990). This difference is due, in part, to the fact that African-American men tend to be diagnosed at later stages of the disease. But, even within stages, survival rates are lower for African-Americans.

African-American men have considerably higher incidence rates (209.6 cases per 100,000 African-American men in 1991) than white men (159.2 cases per 100,000 white men in 1991).

African-American men may have the highest rate of prostate cancer incidence in the world. In addition, their prostate cancer mortality rate is twice as high as the rate for white Americans. In 1991, mortality rates were 24.7 cases per 100,000 white men, and 55.1 cases per 100,000 African-American men. Mortality rates also are increasing much more rapidly among African-American men (about 1.8 percent annually from 1973 to 1991) than among whites (about 1.0 percent annually).

The causes of higher rates of prostate cancer among African-American males are largely unknown. An NCI study found that even when income and education are controlled for, African-Americans have much higher rates than whites. An NCI-supported study is being conducted in three areas of the United States to investigate the reasons for African-Americans' high rates of prostate cancer and other cancers. This case-control study will examine the impact of a wide variety of potential risk factors, including dietary and other lifestyle differences, occupational exposures, and hormonal and genetic differences.

Courtesy National Cancer Institute (NCI)

What does the stage and grade of prostate cancer mean?

Grading and staging prostate cancer is a determination of how fast the cancer is growing and the potential threat to life, and how far it has grown, helping to determine what is the most appropriate treatment for the patient.

Grading is performed by the pathologist who examines the prostate tissue from the biopsy under the microscope examining the appearance of the cancer. The most common grading system is the Gleason Scale which scores the cancer between 5 to 10. The Gleason score is made up of 2 numbers. The first number refers to the grade of the most prevalent cancer cells and the second number refers to the grade of the next most prevalent area.

In general, it is uncommon to have Gleason scores of 5 and less. These indicate very slow growing cancers that are unlikely to spread. Gleason score 6 cancers are slow growing, but there is an 18-30% risk of dying from prostate cancer in 15 years. A score of 7 is an intermediate risk cancer, and cancers with a Gleason score between 8-10 are highly aggressive cancers with a high likelihood of spread and 60-87% risk of death from prostate cancer over 15 years.

Staging of prostate cancer is indication of the size and extent of spread. The prostate examination, PSA, and Gleason scores all give an indication of the likelihood of whether the cancer is locally advanced or metastatic (spread elsewhere). Depending on these tests, it may be reasonable to perform additional imaging studies such as a bone scan (this is used to check whether the prostate cancer has spread to bones which is the most common area of metastatic disease), CT scan (computerized tomography CT) which provides x-ray images of the internal organs of the abdomen and pelvis, or endorectal MRI (magnetic resonance imaging) which may help to indicate whether the cancer is localized or spread outside the prostate capsule.

Staging of Prostate Cancer

What is staging of prostate cancer?

When prostate cancer is diagnosed, tests will be performed to determine how much cancer is present, and if the cancer has spread from the prostate to other parts of the body. This is called staging, and is an important step toward planning a treatment program.

As defined by the National Cancer Institute (NCI), the stages of prostate cancer include the following:

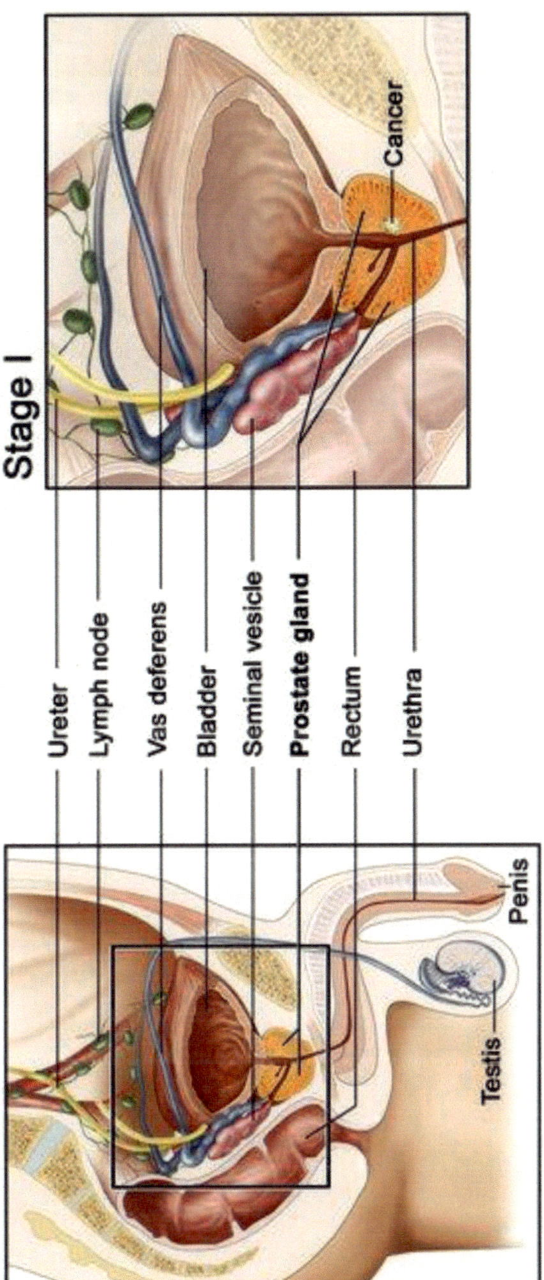
Stage I
Ureter
Lymph node
Vas deferens
Bladder
Seminal vesicle
Prostate gland
Rectum
Urethra
Cancer
Penis
Testis

Stage II

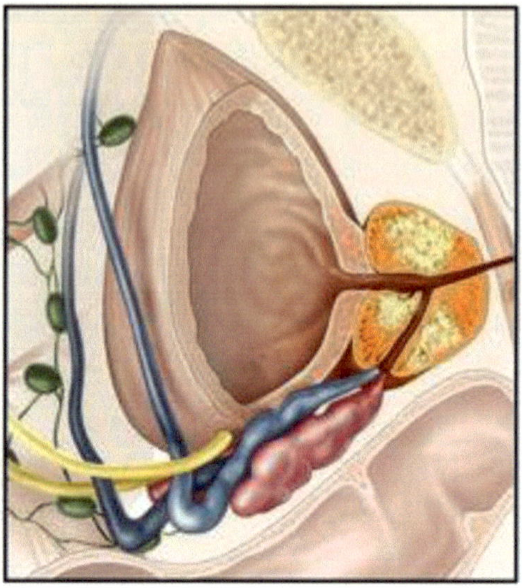

Stage III

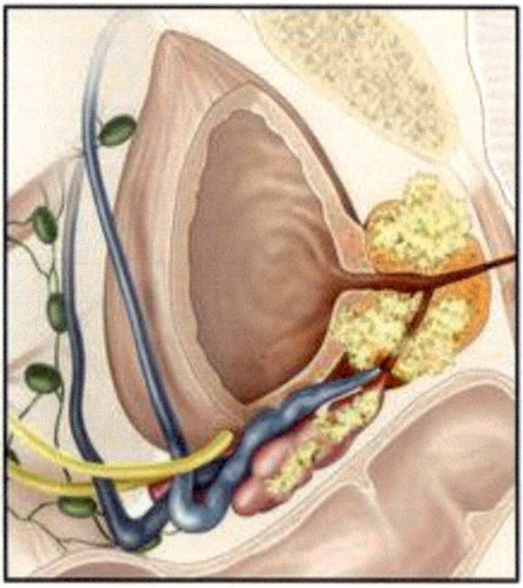

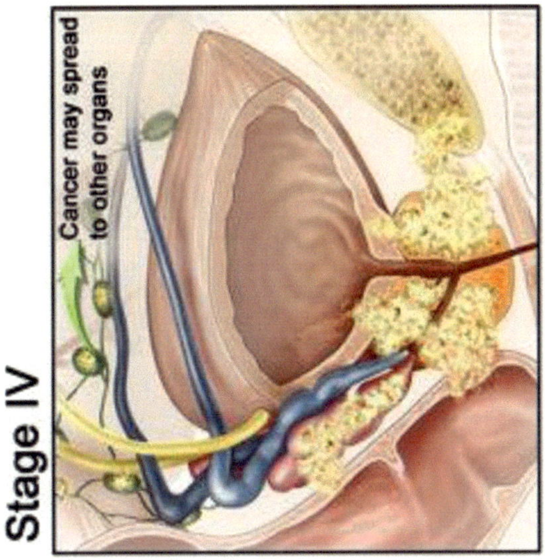

Stage I

- Tumor cells are found in less than 5 percent of prostate tissue removed, and the cells are not very aggressive in nature.
- Sometimes referred to as stage A.

Stage II

- Tumor cells are found in less than 5 percent of prostate tissue removed, and the cells are more aggressive in nature or the tumor is larger in size, but is confined to the prostate gland.
- Sometimes referred to as Stage B.

Stage III

- The tumor has grown through the capsule which surrounds the prostate gland, and may involve seminal vesicles (tubes that carry sperm).

	• Sometimes referred to as Stage C.
Stage IV	• The tumor has spread to other structures beyond the seminal vesicles to any other organ or structure. • Sometimes referred to as Stage D1 or D2.
Recurrent	• The cancer has come back (recurred) after treatment; it may recur in the prostate or in another part of the body. • Sometimes referred to as Stage D3.

Courtesy National Cancer Institute (NCI)

Incidence & Mortality

From 2002-2006, the median age at diagnosis for cancer of the prostate was 68 years of age. Approximately 0.0% were diagnosed under age 20; 0.0% between 20 and 34; 0.6% between 35 and 44; 8.7% between 45 and 54; 29.0% between 55 and 64; 35.6% between 65 and 74; 21.4% between 75 and 84; and 4.7% 85+ years of age.

The age-adjusted incidence rate was 159.3 per 100,000 men per year. These rates are based on cases diagnosed in 2002-2006 from 17 SEER geographic areas.

U.S. Mortality

From 2002-2006, the median age at death for cancer of the prostate was 80 years of age. Approximately 0.0% died under age 20; 0.0% between 20 and 34; 0.1% between 35 and 44; 1.4% between 45 and 54; 7.2% between 55 and 64; 20.1% between 65 and 74; 40.9% between 75 and 84; and 30.3% 85+ years of age.

The age-adjusted death rate was 25.6 per 100,000 men per year. These rates are based on patients who died in 2002-2006 in the U.S.

Incidence Rates by Race

Race/Ethnicity	Male
All Races	159.3 per 100,000 men
White	153.0 per 100,000 men
Black	239.8 per 100,000 men
Asian/Pacific Islander	91.1 per 100,000 men
American Indian/Alaska Native [a]	76.1 per 100,000 men
Hispanic [b]	133.4 per 100,000 men

Trends in Rates

Trends in rates can be described in many ways. Information for trends over a fixed period of time, for example 1996-2006, can be evaluated by the annual percentage change (APC) (See Fast Stats for trends over fixed time intervals). If there is a negative sign before the number, the trend is a decrease; otherwise it is an increase. If there is an asterisk after the APC, then the trend was significant, that is, one believes that it is beyond chance, i.e. 95% sure, that the increase or decrease is real over the period 1996-2006. If the trend is not significant, the trend is usually reported as stable or level.

Death Rates by Race

Race/Ethnicity	Male
All Races	25.6 per 100,000 men
White	23.6 per 100,000 men
Black	56.3 per 100,000 men
Asian/Pacific Islander	10.6 per 100,000 men
American Indian/Alaska Native [a]	20.0 per 100,000 men
Hispanic [b]	19.6 per 100,000 men

Join Point Analyses

A statistical model for characterizing cancer trends which uses statistical criteria to determine how many times and when the trends in incidence or mortality rates have changed. The results of join point are given as calendar year ranges, and the annual percent change (APC) in the rates over each period can be used over a long period of time to evaluate when changes in the trend have occurred, along with the APC which shows how much the trend has changed between each of the join points.

The join point trend in SEER cancer incidence with associated APC(%) for cancer of the prostate between 1975-2006

All Races

Male

Trend	Period
2.6*	1975-1988
16.4*	1988-1992
-11.4*	1992-1995
1.8	1995-2001
-3.4*	2001-2006

The join point trend in U.S. cancer mortality with associated APC(%) for cancer of the prostate between 1975-2006

All Races

Male

Trend	Period
0.9*	1975-1987
3.0*	1987-1991
-0.6	1991-1994
-4.1*	1994-2006

Survival & Stage

Survival Rates

Cancer survival is measured in a number of different ways depending on the intended purpose. Several examples of survival statistics are listed below. They can be calculated by different methods for different purposes. The survival rates presented here are based on the relative survival rate.

A measure of net survival that is calculated by comparing observed (overall) survival with expected survival from a comparable set of people that do not have cancer to measure the excess mortality that is associated with a cancer diagnosis, which measures the survival of the cancer patients in comparison to the general population to estimate the effect of cancer. The overall 5-year relative survival rate for 1999-2005 from 17 SEER geographic areas was 99.7%. Five-year relative survival rates by race were: 99.9% for white men; 96.5% for black men.

The Stage Distribution

Stage provides a measure of disease progression, detailing the degree to which the cancer has advanced. Two methods commonly used to determine stage are AJCC and SEER Summary Stage. The AJCC method (see Collaborative Staging Method) is more commonly used in the clinical settings, while SEER has strived to provide consistent definitions over time with their Local/Regional/Distant staging. Summary Stage 2000 shows that 80% of prostate cancer cases are diagnosed while the cancer is still confined to the primary site (localized stage); 12% are diagnosed after the cancer has spread to regional lymph nodes or directly beyond the primary site; 4% are diagnosed after the cancer has already metastasized (distant stage); and for the remaining 3%, the staging information was unknown. The corresponding 5-year relative survival rates were: 100.0% for localized; 100.0% for regional; 30.6% for distant; and 75.5% for unstaged.

Lifetime Risk

Based on rates from 2004-2006, 15.90% of men born today will be diagnosed with cancer of the prostate at some time during their lifetime. This number can also be expressed as 1 in 6 men will be diagnosed with cancer of the prostate during their lifetime. These statistics are called the lifetime risk.

Lifetime risk may also be discussed in terms of the probability of developing or of dying from cancer. Based on cancer rates from 2004 to 2006, it was estimated that men had about a 44 percent chance of developing cancer in their lifetimes, while women had about a 38 percent chance of developing cancer. Sometimes it is more useful to look at the probability of developing cancer. For example, 8.08% of men will develop cancer of the prostate between their 50th and 70th birthdays.

Prevalence

On January 1, 2006, in the United States there were approximately 2,177,975 men alive who had a history of cancer of the prostate. This includes any person alive on January 1, 2006 who had been diagnosed with cancer of the prostate at any point prior to January 1, 2006, and includes persons with active disease and those who are cured of their disease.

Most prevalence data in SEER is for limited duration because information on cases diagnosed before 1973 is not generally available. It can also be expressed as a percentage and it can also be calculated for a specific amount of time prior to January 1, 2006, such as diagnosed within 5 years of January 1, 2006.

What are the symptoms of prostate cancer?

Prostate cancer in its early stages usually causes no symptoms. The urinary symptoms associated with prostate disease, including poor flow, urinary hesitancy or intermittency, and increased urinary day or night frequency, are more commonly caused by benign prostatic hyperplasia rather than prostate cancer.

Prostate cancer can be detected in its early stages by a prostate examination and the PSA (prostate specific antigen) test.

What is the PSA test?

PSA has been used as a screening test for prostate cancer for almost 25 years. It is a very powerful tool for clinicians in the management of men with prostate cancer with respect to diagnosis, prognosis and monitoring treatment.

It is a sensitive test but it lacks specificity being able to be raised in benign prostatic disease, prostatitis recent ejaculation, as well as prostate cancer, hence its controversy as a screening tool.

The Urological Society of Australasia believes that men in the 50 to 70 year age group with at least a 10 year life expectancy should be able to be screened after appropriate counseling regarding the potential risks and benefits.

At present, there are two schools of thought about PSA testing for prostate cancer. Advocates cite the large number of patients who die of the disease as sufficient justification to initiate screening programs, arguing that lives can be saved by early detection and treatment. Their case has recently been strengthened by evidence for the following:

- the so called latent cancer or ‘autopsy' cases do not appear to be detected by PSA screening
- watchful waiting in younger patients with less differentiated tumors has a high mortality over time
- modern treatments give long term control of most tumors detected by screening and produce fewer side effects than earlier techniques

- the falling death rate from prostate cancer may be attributable to screening

Screening is most likely to benefit patients who have at least 10 years of life expectancy (and inappropriate screening in the much older age group can thus be avoided).

Critics of prostate cancer screening argue that the associated complications and costs are not sufficient to justify widespread implementation of aggressive screening programs. They point out that there has yet to be a randomized trial in any population proving that PSA screening programs prolong life. However, in a recent 10 years analysis of 695 men with localized prostate cancer who were randomly assigned to watchful waiting or radical prostatectomy, radical prostatectomy was shown to reduce disease-specific mortality, overall mortality, and the risks of metastasis and local progression.

Until better data become available, such as trends in prostate cancer mortality and results from ongoing clinical trials, the debate about screening for prostate cancer cannot be resolved.

Only 25% of men who have a PSA level in the suspicious range (i.e. between 4 and 10 ng/mL) will have cancers detected on biopsy.

- If the PSA level is greater than 10 ng/mL, 50% of men will already be incurable.
- 5% of men with prostate cancer have a PSA level below 4 ng/mL (that is, cancers that can be detected only by prostate examination).

PSA changes 10-15% from day to day due to many reasons, including physical and sexual activity but also patient hydration. When interpreting PSA changes, it is important to focus on large changes or consistent rises, PSA increase of >2.0 ng/ml per year or doubling times of two years or less generally reflect clinically significant malignancies. In the PLCO Screening Trial with a database of nearly 40,000 men screened for prostate cancer, a PSA increase of 2.0 ng/mL in the year prior to diagnosis was associated with high-grade (Gleason score > 7) disease on biopsy in 44%. In general terms, a rise of >0.75ng/ml per year is significant.

Transrectal Ultrasound and Prostate Biopsy

If the PSA is elevated or increasing rapidly or there is an abnormal prostate examination, then a transrectal ultrasound and prostate biopsy may be indicated to obtain tissue to make the diagnosis of prostate cancer.

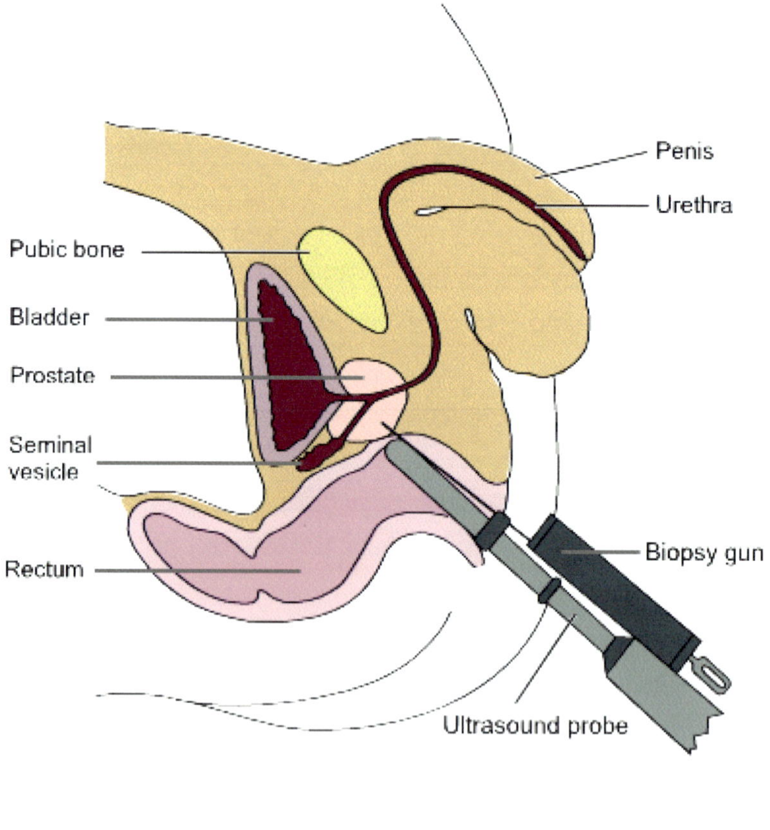

This is performed under intravenous sedation and under antibiotic cover. Ultrasound gives us important information regarding the size and shape of the prostate and prostate abnormalities suggestive of malignancy and an accurate method of sequentially sampling the prostate. We routinely sample 14-20 biopsies depending on the volume of the prostate which increases the diagnostic power of the procedure.

Kegel Exercises

Where do I start?

These exercises take time to master and perfect. At first, you may need to perform these exercises while sitting. As the muscles strengthen, you can progress to exercise standing up. You can try to activate these muscles when performing daily activities or when watching TV (do them every time you see a TV advertisement).

How do I perform the exercises?

1. Squeezing/tightening and drawing in and up around both your anus (back passage) and urethra (bladder outlet). Lift up inside and try to hold this contraction strongly for as long as you can (1-5 seconds). Keep breathing! Now release and relax.

2. Rest 10-15 seconds - repeat Step 1, and remember it is important to rest. Try to hold longer and repeat as many as you are able. Work towards 10 long, strong holds.

3. Now try 5-10 short, fast strong contractions. Do NOT hold your breath, do NOT push down. Do NOT pull your tummy in tightly or tighten your buttocks and thighs.

Anticipate that improvement in pelvic floor muscle strength will take 3-6 months of regular training of the muscles.

The best results will be achieved by seeking help from a physical therapist (with training in continence) who will design an individual training program especially suited to you.

Links

The following listing offers detailed information and services for those who have prostate cancer at any stage:

Prostate Cancer sites

Cancer Council: www.cancercouncil.com.au
Prostate Cancer Foundation: www.prostate.org.au
Lions Australia Prostate Cancer site: www.prostatehealth.org.au

Booklets on Cancer of the Prostate

Localized Prostate Cancer Booklet:
www.prostatehealth.org.au/resources/BI266_LocProsCancBookWeb.pdf
The Cancer Council Booklet:
www.cancercouncil.com.au/editorial.asp?pageid=85

Robot Radical Prostatectomy sites

Intuitive: www.davinciprostatectomy.com
Robot Video: www.davinciprostatectomy.com/video.html

Brachytherapy sites

www.melbourneprostate.org
www.emedicine article on Brachytherapy
www.oncura.com/patients.html

High Intensity Focused Ultrasound Sites

www.melbournehifu.com.au

Information on all Urological Conditions Sites

www.urologyhealth.org

Mens Health site:

www.andrologyaustralia.org

Cancer support sites

Can Survive: www.can-survive.org

Virtual Cancer Centre: www.virtualcancercentre.com

Management of Cancer of the Prostate sites

Hormone Therapy: www.astrazeneca.com.au

Incontinence: www.contfound.org.au

Male Pads: www.tenaformen.com.au

Continence Products: www.paperpak.com

Erectile Disfunction

www.infertility-male.com/erectdys

Professional Organizations

American Urological Association: www.auanet.org

Urological Society of Australasia: www.urosoc.org.au

About the Surgeon

Dr. Sijo Parekattil is Director of Robotic Surgery at Winter Haven Hospital, in the Department of Urology at the University of Florida. He is also the Director of Robotic Microsurgery & Male Infertility at University of Florida. He is an Assistant Professor of Urology and an Adjunct Professor of Bioengineering. Prior to his medical training, he was an Electrical Engineer (University of Michigan), He completed his urological residency training at Albany Medical Center and then went onto complete dual fellowship training from the Cleveland Clinic Foundation, Cleveland, Ohio in Advanced Laparoscopy/Robotic Surgery and Microsurgery/Male Infertility. With a driving interest in surgical techniques and a degree in medicine, his skills have allowed him to incorporate state-of-the art technology, in the field of robotics and microsurgery.

Dr. Parekattil has received numerous awards, including two Annual Innovator Awards from the Cleveland Clinic Foundation. He has published articles in the field of robotic microsurgery and is a co-editor of an upcoming textbook on male infertility with over 50 authors from around the world.

He serves as a Visiting Professor at the University of Zambia, Lusaka, Africa and is involved in a multi-institutional program to deliver medical equipment, training and supplies to the Urology Department at the University Teaching Hospital in Lusaka.

In an effort to stay on the edge of innovative surgical procedures, Dr. Parekattil has performed pioneering work in the arena of robotic surgery, having successfully completed over 600 robotic surgery procedures. In addition, he has lead the development of a multi-disciplinary program dedicated to the treatment of male infertility and groin/testicular pain, which is a major achievement in the field of urology.

Without a doubt, Dr. Parekattil is a world leader in medicine.

Sijo J. Parekattil, M.D.
University of Florida
Director of Robotic Surgery – Winter Haven Hospital
Director of Male Infertility and Microsurgery
Assistant Professor Department of Urology
Adjunct Professor of Bioengineering

Global Urological Institute
Winter Haven Hospital & University of Florida
200 Avenue F NE
Winter Haven, FL 33881
Phone: (863) 293-1121
Email: sijo.parekattil@urology.ufl.edu
Email: sijo_p@hotmail.com
www.urology.ufl.edu
www.roboticinfertility.com
OR
1600 SW Archer Rd,
Gainesville, FL 32610-0247
(352)-273-6815
Fax: (352)-273-7515

Made in the USA
Las Vegas, NV
03 November 2026

57420897R00064